Miscanthus for Bioenergy Production

Miscanthus has been enthusiastically promoted as a second generation biomass crop, and this book provides a comprehensive review of this knowledge.

Miscanthus, also known as elephant grass, is a high yielding grass crop that grows over three metres tall, resembles bamboo and produces a crop every year without the need for replanting or fertiliser application. The rapid growth, low mineral content, and high biomass yield of *Miscanthus* increasingly make it a favourite choice as a biofuel, outperforming switchgrass and other alternatives. There is over 20 years of research evidence to support its promotion as a second generation biomass crop. The author reviews many field measurements of yields as well as the physiology of the crop, and why it is so productive while at the same time requiring low inputs to grow it. It also shows how as a key biofuel crop it can contribute to mitigating climate change and how uptake of the adoption of *Miscanthus* production can be promoted, particularly in Europe and North America.

The book will be key reading for students taking courses in the areas of Environmental Science and Engineering, Climate Change Impacts, Renewable Energy and Energy Conservation. It will also be of interest to researchers of second generation biomass crops, and policy developers working in biofuel production and utilisation.

Michael B. Jones is Emeritus Professor of Botany at Trinity College Dublin, The University of Dublin, Ireland.

Routledge Studies in Bioenergy

https://www.routledge.com/Routledge-Studies-in-Bioenergy/book-series/
RSBIOEN.

Miscanthus for Bioenergy Production

Crop Production, Utilisation and Climate Change Mitigation

Michael B. Jones

Routledge
Taylor & Francis Group

LONDON AND NEW YORK

First published 2020
by Routledge
2 Park Square, Milton Park, Abingdon, Oxon OX14 4RN

and by Routledge
605 Third Avenue, New York, NY 10017

First issued in paperback 2021

Routledge is an imprint of the Taylor & Francis Group, an informa business

British Library Cataloguing in Publication Data
A catalogue record for this book is available from the British Library

Library of Congress Cataloging-in-Publication Data
Names: Jones, M. B. (Michael B.), 1946- author.
Title: Miscanthus for bioenergy production : crop production, utilization and
    climate change mitigation / Michael B. Jones.
Description: Milton Park, Abingdon, Oxon ; New York, NY : Routledge,
    2019. | Series: Routledge studies in bioenergy production | Includes
    bibliographical references and index.
Identifiers: LCCN 2019004462 (print) | LCCN 2019005418 (ebook) | ISBN
    9781315108155 (eBook) | ISBN 9781138091245 (hbk) | ISBN 9781315108155
    (ebk)
Subjects: LCSH: Miscanthus. | Energy crops.
Classification: LCC SB197 (ebook) | LCC SB197 .J66 2019 (print) | DDC
    633–dc23
LC record available at https://lccn.loc.gov/2019004462

ISBN 13: 978-0-367-78757-8 (pbk)
ISBN 13: 978-1-138-09124-5 (hbk)

Typeset in Sabon
by Taylor & Francis Books

Contents

Illustrations

Figures

Tables

Preface

Miscanthus for Bioenergy Production – Crop Production, Utilisation and Climate Change Mitigation

At present, fossil fuels supply more than 80% of the world's energy demand but they also release about 37 Gt of the greenhouse gas CO_2 annually to the atmosphere, and this is largely responsible for global warming and climate change. Bioenergy, unlike fossil fuels, is a renewable energy source that has the potential to move the planet into a more sustainable future because the CO_2 released has been recycled through photosynthesis, dramatically reducing the net release of greenhouse gases. Several scenarios indicate that bioenergy will have a 25% share of the global primary energy supply by 2050. It will also transform the way we use our land and natural resources, mainly by exploiting inefficiently used land, extensive pastures, degraded land and excess agricultural capacity and residues.

In order to achieve these ambitious targets there is a need to dramatically increase biomass production for energy on a global scale. A major route for achieving this is to select high-yielding bioenergy crops that are more efficient in their use of light energy, water and nutrients, as well as being more resilient to climate change, and develop market-based incentives for growing and utilising the biofuels.

Bioenergy production can be economically beneficial for rural communities, contributing to farm incomes and increasing rural employment. Also, when properly managed, energy crops can have positive impacts on the environment, helping to maintain soil quality and increasing carbon accumulation and improving water quality. Although, negative impacts on biodiversity and water quality are a danger if not managed effectively. The use of advanced biofuels can help realise the maximum potential of bioenergy with the least negative impacts.

Bioenergy can be supplied by number of sources but the topic of this book is a bioenergy crop most commonly referred to by the genus name, Miscanthus. The genus Miscanthus comprises about a dozen grass species that are closely related to sugarcane. These grasses are perennial and rhizomatous and also they have a particularly efficient form of photosynthesis.

The history of research on *Miscanthus* as a biofuel goes back to the 1970s in Europe where it was included as part of an EU Framework Programme on 'biomass as a source of bioenergy'. Much of the early work was carried out under the Agro-Industry Research (AIR) programme of the EU Directorate General for Agriculture, which in 1992 established the *Miscanthus* Productivity Network. The network had 17 partners located in ten countries throughout the Europe and Mary Walsh and I edited a book that reviewed the state-of-the-art of *Miscanthus* in Europe that was written by members of the network and was published in 2001 (Miscanthus *for Energy and Fibre*, published by James and James).

Since the turn of the century, interest in and research on *Miscanthus* has blossomed. In 2001, publications dealing with *Miscanthus* (cited in Web of Science) were only 50, however, over the following years they rose exponentially to more than 500 per year in 2013, and since then they have run at between 550 and 600 per year. This book aims to review the startling progress in research output on *Miscanthus* since 2001 and assess how this information is being applied to establish *Miscanthus* as the bioenergy crop of choice for farmers, particularly in Europe and North America. It should also provide a reference for students and early career researchers who are excited by the possibility of energy crops contributing to the mix of solutions to our rapid transition to renewable energy. Finally, as scientists we are increasingly encouraged to engage productively with people who make policy. I hope therefore that this book can provide the type of informed summary that policy makers can incorporate into their policy development for a rapidly evolving landscape where hard decisions will need to be taken to avoid a future blighted by the extremes of a changing climate.

I would like to express my thanks to my fellow researchers, and in particular the many post-doctoral and post-graduate students in Trinity College, who have accompanied me on my journeys with 'energy crops' and 'C4 grasses and sedges'. I also owe an especial debt to my wife, Sue, who has encouraged me throughout my career and especially in the writing of this book.

Finally, I think the following quote from Jonathan Swift, 1667–1745, Dean of St Patrick's Cathedral, Dublin, is particularly appropriate for our scientific community striving to provide a better world for future generations, including my grandsons, Daniel and Rory:

'Whoever could make two ears of corn, or two blades of grass, to grow upon a spot of ground where only one grew before, would deserve better of mankind, and do more essential service to his country, than the whole race of politicians put together.'

Glossary and abbreviations

Agroforestry (AF) land use management system in which trees are grown among crops or pastureland.

Biodiesel An oil based biofuel, typically produced from vegetable fats, such as rapeseed, sunflower seed, soya bean and palm oil, and blended with conventional diesel for use in motor vehicles.

Biodiversity The variety of different life forms in a given area. High biodiversity is viewed as an indication of a healthy ecosystem.

Bioeconomy The parts of the economy that use renewable biological resources to produce food, energy and industrial goods.

Bioenergy Energy from biomass with most common applications in the transport, heat and electricity sectors.

Bioenergy with carbon capture and storage (BECCS) GHG mitigation technology which combines bioenergy use with geologic carbon capture and storage.

Bioethanol An alcohol based biofuel, typically produced from starch and sugar crops, such as wheat, corn, barley and sugar beet or cane, and blended with petrol for use in motor vehicles.

Biofuel A fuel produced from biomass. The two most common types of biofuel are bioethanol and biodiesel. Biofuels are also distinguished by the type of feedstock from which they are produced as first, second and third generation:

– First generation biofuels (also referred to as 'conventional' biofuels): Biofuels produced from food or animal crops.

– Second generation biofuels (also referred to as 'advanced' biofuels): Biofuels derived from dedicated energy crops (e.g. *Miscanthus*, switchgrass, short rotation coppice and other lignocellulosic plants), agricultural residues, forest and sawmill residues, wood wastes and other waste materials (e.g. used cooking oil and municipal solid waste). A key characteristic is that these feedstocks cannot be used for food.

– Third generation biofuels (also referred to as 'advanced' biofuels): Biofuel produced from microalgae through conventional transesterification or hydro-treatment of algal oil.

Carbon capture and storage (CCS)   The process of capturing waste CO_2 from large point sources, transporting it to a storage site and depositing where it will not enter the atmosphere.

Carbon dioxide removal (CDR)   Refers to a number of technologies where the objective is to remove CO_2 from the atmosphere e.g. Biochar, enhanced weathering, CCS, direct air capture, soil carbon sequestration.

Carbon footprint   Total life cycle emissions of greenhouse gasses from a system, expressed in carbon dioxide equivalents (CO_2 eq.). For biofuels, the life cycle typically includes cultivation and harvesting or collection of feedstocks (as relevant), their processing, production and use of biofuels, waste management and all intermediate transportation steps.

Combined heat and power (CHP)   The simultaneous generation of useable heat and electricity in a single process. Considered to be a highly efficient energy production process.

Cradle to gate   Life cycle stages from the extraction of raw materials ('cradle') to the point at which the product leaves the production facility ('gate').

Cradle to grave   Life cycle stages of raw materials ('cradle') to final disposal of waste ('grave').

Dedicated energy crops   As opposed to some crops that can be used in the production of biofuels but also have food and feed markets, dedicated energy crops are grown especially for the purpose of producing heat, electricity or transport biofuels. Dedicated energy crops are non-food crops, including *Miscanthus*, switchgrass, short rotation coppice and other lignocellulosic plants, as well as non-food cellulosic material, except saw logs and veneer logs which have an alternative market outlet.

Ecosystem services   Environmental resources used by humans, including clean air, water, food and materials. According to the Millennium Ecosystem Assessment, ecosystem services can be classified into: (i) supporting services (e.g. photosynthesis, nutrient and water cycling), (ii) provisioning services (e.g. food, water), (iii) regulating services (e.g. air quality, climate etc. regulation) and (iv) cultural services (e.g. recreational amenities and aesthetic value landscape).

EU Biofuels Directive   Originally Directive 2003/30/EC, later amended by Directive 2009/28/EC (see 'EU Renewable Energy Directive'). It stipulated implementation of national measures by member states aimed at replacing 5.75% of all transport fossil fuels (petrol and diesel) with biofuels.

EU Fuels Quality Directive (FQD)   Directive 98/70/EC (as amended), requiring suppliers to reduce the life-cycle greenhouse gas intensity of transport fuels and introducing sustainability criteria for biofuels.

EU ILUC Directive   Directive 2015/1513 amends the Renewable Energy Directive and the Fuel Quality Directive to take account of the effect of indirect land-use change (ILUC) and aims to encourage the transition away from first generation biofuels.

EU Renewable Energy Directive (RED)   Directive 2009/28/EC of the European Parliament and of the Council of 23 April 2009 on the promotion of the use of energy from renewable sources and amending and subsequently repealing Directives 2001/77/EC and 2003/30/EC. The RED requires member states to ensure that 10% of the energy used in transport is from renewable sources by 2020.

Functional unit A measure of the function of the system of interest and used as a unit of analysis in life cycle assessment (LCA). For example, 1MJ (10^6 joules) is often used as the functional unit in LCA studies of biofuels, reflecting its main function as the provision of transportation energy.

Global warming potential (GWP) Cumulative radiative forcing from the emission of a unit mass of a greenhouse gas relative to carbon dioxide. The radiative forcing effect is integrated over a period of time, 20, 100, or 500 years, with 100 years being used most often. GWP is expressed in carbon dioxide equivalents (CO_2 eq.) where GWP of one mass unit of CO_2 is equal to one. In life cycle assessment, the term GWP is used to denote the climate change impact from the total emissions of greenhouse gas from a system and is often used interchangeably with the term 'carbon footprint'.

Greenhouse gases (GHG) Gases in the atmosphere that absorb and re-emit infrared radiation reflected from the Earth, This causes the so-called 'greenhouse effect' whereby heat is trapped in the atmosphere making the Earth warmer and leading to climate change. The basket of six GHGs produced by human activities are: carbon dioxide (CO_2), methane (CH_4), nitrous oxide (N_2O), hydrofluorocarbons (HFCs), perfluorocarbons (PFCs) and sulphur hexafluoride (SF_6).

Integrated assessment models (IAMS) Modelling used by the environmental sciences and environmental policy analysts. They integrate knowledge from different academic disciplines.

Land-use change (LUC) Change in the purpose for which land is used by humans (e.g. crop-land, grassland, forest-land, wetland, industrial land). Two types of LUC are distinguished:
– Direct LUC: Change in the use of land at the location of production of the crop of interest.
– Indirect LUC (iLUC): Change in the use of land elsewhere occurring indirectly as a result of displaced demand previously destined for food, feed and/or fibre markets owing to biofuel demand.

Leaf area index (LAI) The projected area of leaves over a unit of ground surface area.

Life cycle assessment (LCA) A method used to quantify environmental impacts of products, technologies or services on a life-cycle basis. An LCA study can be from 'cradle to grave' or from 'cradle to gate'. A 'cradle to grave' study of a product considers all life-cycle stages from extraction of materials and fuels ('cradle') through production and use of the product to its final disposal as waste ('grave'). A 'cradle to gate' study does not follow the product to the use stage but stops at the factory 'gate' where the product has been produced.

Marginal land   Degraded or generally poor quality land, unsuitable for agricultural, housing and other uses.

Natural capital   Natural resources such as soil, air, water and all living things. They provide a wide range of services, often called ecosystem services, which contribute to human well-being (see also 'ecosystem services').

Negative Emissions Technologies (NETS)   Negative emissions mean reducing the amount of CO_2 in the atmosphere by capturing it and storing it in a safe place.

Nitrogen use efficiency (NUE)   The ratio of increase in plant biomass to increase in plant nitrogen content over the growing season.

Perennial rhizomatous grasses (PRGs)   Long-lived grasses that regrow each year from their rhizomes.

Photosynthetic nitrogen use efficiency (PNUE)   The ratio of net leaf photosynthesis in full sunlight to the leaf nitrogen content.

Radiation use efficiency (RUE)   The crop biomass produced per unit of radiation intercepted by the canopy.

Renewable fuel standard (RFS)   An American federal programme that requires transportation fuel to contain a minimum proportion of renewable fuel.

Short rotation forestry (SRF)   The practice of cutting fast growing trees that are repeatedly harvested every 4–10 years.

Soil carbon sequestration (SCS)   Long-term storage of carbon in the soil.

Soil organic carbon (SOC)   Carbon present in soils as organic matter. It includes carbon in living plants and in materials derived from plant remains, such as humus and charcoal.

Supply chain   The whole production chain from the production of feedstock for the production of biofuels up to the biofuel user or trader.

Sustainable Development Goals (SDGs)   The 2030 Agenda for Sustainable Development adopted by the UN member states in 2015. There are 17 SDGs.

System boundary   The boundary drawn around the system of interest, denoting unit processes and stages in the life cycle considered in a life cycle assessment study.

System expansion or substitution   Applied in life cycle assessment studies to estimate environmental impacts of the product of interest produced in a system that also co-produces other products. The system boundary is expanded to consider alternative ways of producing the co-products. The system is 'credited' for displacing (substituting) the need for these alternative production systems by subtracting their impacts from the overall impacts of the system under study.

Verification   The process of providing assurance of biofuel sustainability data or other fuel-related data (e.g. place of purchase, volume produced) supplied on behalf of reporting parties. Verifiers must be independent of the reporting party whose data they are verifying.

Water use efficiency (WUE)   The water use efficiency of productivity is the ratio of biomass produced to the amount of evapotranspiration.

1 Bioenergy and its global potential

Introduction

Bioenergy is a term used to describe any kind of renewable energy generated from material derived from recently living organisms which includes any plants, animals and their by-products. Bioenergy use today is being increasingly expanded to provide not only sustainable alternatives to fossil fuels but also additional income for rural communities where jobs are generated through feedstock production and harvesting on farms, and in transport and processing (EASAC, 2012).

Before the rapid utilisation of fossil fuels started in the 17th century humankind was totally reliant on renewable energy for cooking and heating. Plant biomass in the form of peat, wood and herbaceous vegetation was burned to provide heat. As demand increased, vast areas of forests were cleared not only for energy but also to provide material for construction of buildings and means of transport; however, removal was not matched by replenishment. In effect, what was initially renewable became non-renewable. However, once the industrial revolution got underway additional sources of energy were utilised and fossil fuels, coal initially and oil later, were recognised as more convenient sources of condensed energy. What was important about the fossil fuels was that there were sufficient reserves of them largely stored underground. Initially there appeared to be unlimited reserves but as demand increased there were constraints on supply and the energy extraction market began to talk about ultimately running out of these supplies as the reserves were depleted and they became increasingly difficult to extract. Other issues also became clear, such as the pollution of the atmosphere as a result of burning these fuels, in particular sulphur dioxide, nitrogen oxides, ozone and particulate matter. Later the realisation that fossil fuels were responsible for the rapidly increasing concentration of carbon dioxide (CO_2) in the atmosphere and that the rise in concentration was responsible for global warming gave increasing urgency to reducing these emissions (Houghton, 2004). We have now reached a situation where the United Nations Intergovernmental Panel on Climate Change (IPCC) Paris Agreement (2015) calls for a total ban on emitting fossil fuel CO_2 and possibly even a requirement to remove CO_2 from the atmosphere in order to prevent warming rising to more than 2°C above pre-industrial levels. Energy production in the future should therefore be carbon neutral or even carbon negative if society is to achieve the Paris targets.

The so called energy crises of the 1970s were one of the first drivers of a search for renewable energy options which could be considered to be close to carbon neutral (i.e. their use does not lead to net greenhouse gases (GHG) emissions to the atmosphere). The 1970s energy crises were a period when the major industrial countries of the world, particularly the United States, Canada, Western Europe, Japan and Australia faced substantial oil shortages, real and perceived, as well as elevated prices. The two worst crises of this period were in 1973 and 1979 when the Yom Kippur War and the Iranian Revolution triggered interruptions in Middle Eastern oil exports.

There are a number of renewable energy options including solar, wind, hydro and biomass. They are not 100% carbon neutral as they require energy inputs in manufacturing solar panels, wind turbines, dams and water turbines. Other particular drawbacks are that wind and solar are episodic producers of energy and, along with biomass, they may require access to land which we currently need for growing crops.

Biomass and biofuels

Of the renewable energy options, biomass is arguably the most flexible. Biomass can provide energy in solid, liquid and gaseous forms (Figure 1.1). Energy can be released by combustion of solid biomass. Combustion of biomass, often burned together with coal, generates electricity and electricity/steam in district heating systems provided for office and apartment buildings.

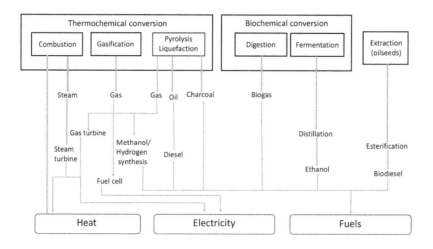

Figure 1.1 Biomass as a source of energy – solids, liquids and gaseous forms (Adapted from Faaij, 2006).

Biomass can also be converted to liquid fuel and biogas, referred to collectively as 'biofuel', which can be used in a range of already developed facilities from single-family units up to industrial scale. Liquid biofuel can be produced by fermentation of sugars derived from biomass to produce bioethanol or by extracting oils mainly from seed to produce biodiesel. Biogas is also produced by fermentation of biomass. Biogas is a mixture of methane and CO_2 produced by anaerobic digestion of biomass raw materials which can be used to generate electricity or can be upgraded to methane to use in transportation.

Of the biomass products, liquid biofuels have attracted most attention in recent years, largely because of a drive to provide energy security when supply of liquid fossil fuels is threatened or too costly. One of the most desirable attributes of biofuels is that they can be easily blended into petroleum fuels or relatively easily used as alternatives to diesel fuels.

Liquid biofuels, primarily bioethanol and biodiesel, support transport needs. In 2010 there were around 800 million cars in the world and by 2050 that number is expected to reach 1.7 to 2.1 billion. By then biofuels could be supplying up to 30% of the demand for liquid fuels. In 2007, 27 out of 50 countries surveyed had policies under consideration or had enacted mandatory requirements for biofuels to be blended with traditional transport fossil fuels, and 40 had legislation to promote biofuels (Smith, 2010). Evidence of the take-off in biofuel use since the turn of the century is that between 2002 and 2006, the amount of land used to grow biofuel crops quadrupled and production tripled (Coyle, 2007).

According to the International Energy Agency (IEA, 2018) the output of transport biofuels, in the form of grain ethanol in North America has risen by 5% year on year to almost 60 billion litres in 2017, but is anticipated to be broadly stable between 2018 and 2023. However in the EU, conventional first generation biofuel is forecast to contract by 10% as the policy framework becomes less supportive of conventional biofuels after 2020. This is because the updated Renewable Energy Directive Framework (EU, 2017) favours advanced (second generation) biofuels over conventional biofuels, increasing the transport sector renewable energy consumption target from 10% to 14% but capping the conventional contribution at 7%. Clearly, higher production of advanced biofuels will be important in the long-term because they generally offer more significant greenhouse gas emission reductions relative to conventional biofuels. Also, as they are produced from non-food feedstocks, they do not compete with food crops for prime agricultural land. Advanced biofuels are manufactured from various types of non-food biomass and are referred to as lignocellulosic or more simply cellulosic biomass. Cellulosic feedstock consists of a mixture of cellulose, hemicellulose and lignin, as well as numerous minor components and minerals. Many of these materials, particularly lignin are much more resistant to chemical breakdown than are sugars. Processing conditions are therefore more severe in temperature, pressure, time and chemical reagents to convert to simple products for fermentation.

Environmental impacts

It has been demonstrated that when properly managed, energy crops can have positive impacts on the environment by, for example, helping to maintain soil quality and increasing carbon accumulation and improving water quality. However, it has been increasingly recognised that, under some circumstances bioenergy development can cause significant negative environmental and social impacts including increased greenhouse gas emissions, deforestation, biodiversity loss, soil erosion, soil carbon loss, unsustainable water use, conflicts over land rights, food shortages and spikes in staple food-crop prices. In fact, some have argued that biomass is inefficient in meeting human demands that could be met in other ways. Certainly, the large numbers of energy inputs that go into the production of bioenergy crops and the extraction of useful energy have made it very difficult to guide policy makers towards identifying the most viable choices and away from dangerous and costly misjudgements. It is therefore necessary to develop research in this area to fully understand the global consequences of large scale bioenergy production. Bioenergy development should be continuously monitored for its environmental and social sustainability to relieve uncertainties that policy makers may have about the cost effectiveness and environmental sustainability of bioenergy production.

Global potential

Despite the concerns and scepticisms expressed by many researchers and analysts, international support for bioenergy production for transport and biomass heat has increased significantly over the last 20 years. For example, the EU has set a target of 50% of total energy consumption to be from renewable resources in 2050, which would amount to 39% from biomass following the current trajectory (European Commission, 2011, 2013). In the USA the Renewable Fuels Standards (RFS), part of the Energy Independence and Security Act (EISA) of 2007, establishes an annual production target of 36 billion gallons of biofuels by 2022 (EISA, 2007). The International Energy Agencies analysis and forecasts (IEA, 2013) reported that half of all renewable energy consumed in 2017 came from bioenergy and they forecasted that up to 2023 around 30% of the growth in renewable consumption would be expected to come from bioenergy. Several scenarios indicate that bioenergy will have a share of 25% of the global primary energy supply by 2050 (IPCC, 2011). In some analyses, bioenergy is seen as providing a bridging technology until the middle of the current century when it most likely will be overtaken by wind and solar (WBGU, 2008; IEA, 2018). As a consequence, bioenergy will transform the way we use our land and natural resources, mainly by exploiting inefficiently used land, extensive pastures, degraded land and excess agricultural capacity and residues.

In order to achieve the high targets for bioenergy, projections from the Intergovernmental Panel on Climate Change (IPCC, 2011) and the International Energy Agencies Biofuel Technology Roadmap (IEA, 2011) both estimate that 500 EJ yr^{-1} of biomass primary energy could potentially be available. However, models used to project future production potential produce a very wide range of results ranging from 27 to 1546 EJ yr^{-1} and unsurprisingly some researchers are cautious about proposals to expand the bioenergy industry based on these very uncertain projections. A major criticism of biomass potential assessments is that there is a lack of standardisation and consistent methodologies. Slade et al. (2014) suggest that the choice of alternative assumptions, rather than methodological differences, are the main reason for the disparity in assessments. They propose that there should be more precision in the use of terms such as abandoned land and surplus forestry waste for biofuel which are vague and should be avoided. Reaching the stated technical energy potential depends on many important factors, including land and water availability, feedbacks between food, livestock and energy systems and climate change (Coelho et al., 2012). The most important parameter assumptions determining the technical potential for energy crops are restrictions on land available for growing the crops, the future productivity of energy crops and the sustainability restrictions on their deployment.

This has led to a reassessment of the assumptions in some of the models and a revised upper-end estimate of potential biomass availability of 40–110 EJ yr^{-1} (Searle and Malins, 2015). These conclusions suggest that many technical projections and aspirational goals for future bioenergy use could be difficult or even impossible to achieve sustainably. There is clearly a concern that over-ambitious policies that promote increasing bioenergy use could result in unmet targets, unsustainable practices and most importantly negative impacts on the world's food supply. Furthermore new and emerging demands on sustainable biomass to supply the burgeoning 'bioeconomy' (see Chapter 6) mean that there will be other calls on the biospheres' renewable supplies.

There are now vociferous camps of supporters and critics of bioenergy. Supporters of bioenergy argue that bioenergy can help to secure energy supply and to mitigate climate change, as well as creating development opportunities in the rural areas of industrialised and developing countries. Critics argue that growing energy crops will heighten land-use conflicts as food cultivation, nature conservation and bioenergy production compete for land, and bioenergy impacts negatively on the climate. However, these are complex issues with no simple rights or wrongs because of the high levels of scientific uncertainty and the multiplicity of interests that are involved. Analyses need to be aware of the many possible bioenergy pathways available and the different characteristics and global linkages among the effects, and that no single sweeping statement is possible.

The two main principles are that first, bioenergy should contribute to mitigating climate change by replacing fossil fuels and secondly, bioenergy should help to overcome energy poverty of the estimated 2.5 billion people who currently have no access to affordable and safe forms of energy, prevent damage to health, and reduce pressures on natural ecosystems. Bioenergy is envisaged to maintain its current position as the highest contributor to global energy in the short term with dedicated energy crops set to provide an increasing proportion of the biomass feedstock in the coming decades. In the more distant future the opportunities for biomass crops include development of bio-refineries and the development of small scale, distributed energy systems (Sims et al., 2006).

Beneficial biofuels

In 2008, when global food prices peaked, biofuels were blamed for starving the poor, disturbing markets, making unsustainable the use of land and water and, especially due to indirect land use change, resulting in negative GHG balances (Dornburg et al., 2010; Smith, 2010). However, Tilman et al. (2009) summarised the position as follows: 'In a world seeking solutions to its energy, environmental, and food challenges, society cannot afford to miss out on the global greenhouse gas emission reductions and the local environmental and social benefits when biofuels are done right. However society also cannot accept the undesirable impacts of biofuels done wrong.' In other words, we need to identify beneficial biofuels that are derived from feedstocks that are produced with much lower greenhouse gas emissions than traditional fossil fuels, with little or no competition with food production, and with as little as possible negative impacts on the land on which they are cultivated.

The search for beneficial biofuels should focus on sustainable biomass feedstocks that neither compete with food crops nor directly or indirectly cause land-clearing and that offer advantages in reducing GHG emissions. To these conditions, Rist et al. (2009) suggested the addition of 'the maximisation of social benefits' which requires an assessment of the potential opportunities and risks to rural communities afforded by biofuel feedstock cultivation. Each biofuel should be evaluated on its net benefits for society based on a full life-cycle analysis that includes, among other factors, its effect on net energy supply, the global food system, GHG emissions, soil carbon and soil fertility, water and air quality, and biodiversity.

Another way in which biofuel production can be assessed is in terms of the ecosystem services that it provides. These are the many and varied benefits that humans freely gain from the natural environment and from properly functioning ecosystems. While scientists and environmentalists have discussed ecosystem services implicitly for decades, the Millennium Ecosystem Assessment (2005) in the early 2000s popularised the concept. The four widely accepted categories of ecosystem services are, supporting, provisioning, regulating, cultural and biodiversity. Supporting services are necessary for the production of all other ecosystem services including primary production, atmospheric O_2, soil formation, nutrient cycling, and water cycling. Provisioning services are the products from ecosystems including fuels and genetic resources, while regulating services are air quality maintenance, GHG mitigation, erosion control, and pollination services. The concept of ecosystem services directly links ecosystem impact and human wellbeing, two key elements of the biofuel debate. An assessment of ecosystem services allows an understanding of the interrelations between ecosystem change and human wellbeing. Although the concept of ecosystem services has been used widely to understand the impacts of numerous human activities on diverse social-ecological systems there is relatively little literature linking biofuel production and ecosystem services. Bioenergy feedstock production has the potential to influence many ecosystem services, although GHG balance of bioenergy crops is widely considered of overriding importance in assessment of sustainable deployment of bioenergy.

The way forward – identifying suitable energy crops

Energy crops can take many different forms and can be utilised in a variety of ways from simple combustion to complex bioconversion processes (Ragauskas et al., 2006). However, there is a strong case that it is in the context of bio-fuels as transportation fuels that energy crops have the biggest opportunity to make an impact (Heaton et al., 2008b). It will also be essential to produce energy crops with minimum resource inputs; in other words the 'resource use efficiency' of production must be maximised. At present, most of the so-called 'first generation' feedstocks for biofuel production are produced from two crops that have, until recently, been used primarily for food production; these are sugarcane (*Saccharum officinarum*) and maize (corn) (*Zea mays*). The two main producers of liquid biofuels today are Brazil (20 billion litres yr^{-1}) and the United States (24 billion litres yr^{-1}) which produce bioethanol from, respectively, the fermentation of sugars extracted from sugarcane or derived from the hydrolysis of starch in maize.

The three distinct goals associated with development of biofuel feedstocks are: maximising the total amount of biomass produced per hectare per year, maintaining sustainability while minimising inputs, and maximising the amount of fuel that can be produced per unit of biomass. It is anticipated that the next (second) generation of feedstock for ethanol production will utilise, in addition to sugars and starch, the cellulose and hemi-celluloses from per-ennial grasses, wood chips and agricultural residues. However this depends on the development of processes which can extract fermentable sugars from the cellulose and hemicelluloses or produce other products such as dimethyl furan for liquid fuels (Somerville et al., 2010).

The grand challenge for biomass production is to develop the second gen-eration energy crops which have a suite of desirable physical and chemical traits to aid bioethanol and other bi-product extraction, while maximising biomass yields. Achieving this will depend on identifying the fundamental constraints on productivity of bioenergy crops and in particular selecting plant life forms which are particularly well-suited to maximising outputs in terms of biomass yield while minimising inputs. It is now widely recognised that perennial rhi-zomatous grasses (PRGs), such as *Miscanthus*, possess these desirable characteristics and it is likely that they will become the dedicated bioenergy crops for the future. The most productive of the PRGs have C$_4$ photosynthesis, a particular type of photosynthesis which uses light, water and nutrient resources very efficiently. Currently, the second generation bioenergy crops are largely undomesticated and have not been subject to centuries of improvement, as have our major food crops. Breeding of appropriate species and genotypes to suit specific climates and soil conditions will be required. The foundation of this approach will be to use the high through-put tools of genomics, metabolomics and phenomics to rapidly develop the understanding needed to develop novel, second generation bioenergy crops. However, it will be important for bioengi-neers using high-throughput techniques to work closely with agronomists and breeders, if the novel genotypes are to be optimised for the multitude of field conditions around the world. Currently, we are only in the preliminary stages of breeding programmes for the leading candidate crops and are unlikely to see significant productivity gains in the near future (Vermerris, 2008; Clifton-Brown et al., 2018).

In the meantime, a prudent development strategy would see fuels developed in phases. This process will allow policy support, institutional capacity and regulatory requirements to develop as required for each phase rather than all at once. The phased approach would allow benefits of biofuels to be achieved at each phase while allowing preparation for the next. The first phase might be the use of existing plant biomass that is not currently utilised but only where this utilisation has very low environmental impact. Careful environmental assessment of this use would need to be carried out before any harvesting could take place. The first phase would be to use the biomass for combustion but later there would be opportunities to move to second generation use for bioethanol production and the development of bioprocessing for higher value products.

Low carbon bioeconomy

Bioeconomy is a term used to describe the biotechnological and life sciences part of the existing economy but it is also used for describing an economy which is predominantly based on biomass rather than fossil-based resources. These resources are 'low carbon' because they are transitioning to renewable sources of energy. Strictly speaking they are not 'low carbon' as they are still carbon based, but they recycle carbon on a short timescale.

The ultimate aim for what is described as a low carbon bioeconomy is to increase the use of low carbon sources (i.e. sustainable biomass) as the feedstock for not only the production of energy but for chemicals and materials. In the not-so-distant past, the world relied almost entirely on renewable resources, including biomass, for food, energy, and shelter. In the future this could be once again true – many modern needs including plastics, materials of construction, clothing and more importantly energy, can be met by biomass. Realising this potential will require the development, demonstration and deployment of a number of innovative processes that can meet strict sustainability criteria. An expanded sustainable bioeconomy, respecting biodiversity, can also provide wider environmental, social and economic benefits by replacing fossil feedstocks, by creating jobs, and by promoting regional development, in alignment with Sustainable Development Goals (SDGs).

Despite the growing consensus, bioenergy and bio-based product deployment and investments are not growing quickly enough, and the technology suffers from a number of barriers, including early stage scale challenges, financial risks, oil and feedstock price volatility and policy uncertainty. This is why creating the conditions for scaling up the low carbon bioeconomy is both an urgent and vital challenge. There is substantial demand for additional biomass to fulfil the global bioeconomy goals. However, there is the possibility that a large increase in demand for biomass could undermine the sustainability of a bio-based economy. Furthermore we have seen that additional land use could have negative impacts, including loss of both biodiversity and soil carbon. Higher crop demand could lead to higher fertiliser use (resulting in higher GHG emissions), increased demand for pesticides, and water and soil pollution. Probably most significant would be increased competition for land-use between food or bioenergy production. In summary, although the technical potential is large, the actual supply of low-cost sustainably produced biomass is limited (Ragauskas et al., 2006).

Conclusions

The extent to which energy crops can deliver sustainable biomass on a global scale remains poorly understood (Slade et al., 2014). Estimates of global production cannot be measured directly, so they are extrapolated from limited country-specific data to obtain approximate global supply. However, the wide range of estimates of biomass potential and the lack of standardised assessment methodologies fosters uncertainties amongst policy-makers. Ultimately, this has given rise to a general sense of unease about the future of biomass and whether it offers realistic opportunities or is a utopian vision that stands little chance of being realised. This is sometimes presented as a 'Jekyll and Hyde' syndrome, taken from Robert Louis Stevenson's *The Strange Case of Dr Jekyll and Mr Hyde* which refers to a character that switches continuously between good and bad. The scepticism has been fuelled by a relatively widespread view from elements in the science and engineering community that enthusiasts for exploiting biomass for energy are totally unrealistic about the potential contribution that this form of energy can make to the overall renewable energy mix. This is best summarised by the view of David MacKay (2008) in his widely acclaimed book *Sustainable Energy – without the hot air* that 'bioenergy is in reality too inefficient to meet human demands and that these demands could be met in other ways because it requires massive quantities of land and water for modest quantities of energy, and 'biofuels made from plants ... can deliver so little power, I think they are scarcely worth talking about'.

However, it might be argued that uncertainties around the potential scale of bioenergy supply are as much political and social as technical (Smith, 2010), so that government policies, which reflect the political ambitions of the government of the day, can switch dramatically with the shade of the dominant political thinking. There have been a number of examples of this in Europe, but perhaps the most significant has been the election of Donald Trump to US President and the policy of protecting coal and other fossil fuel extraction.

In terms of global energy demand, the need is critical. According to recently updated analysis and long-term scenarios by key international agencies, such as the International Energy Agency (IEA, 2018), 'sustainable bioenergy is an indispensable component of the necessary portfolio of low carbon measures, with a high risk of failing to meet long-term climate goals without its contribution'. Bioenergy is seen as being key in several areas, including heating and transport, and particularly in heavy freight, maritime, and air transport, where other practical options are scarce.

In order to limit the increase in global average temperature to well below 2.0°C above preindustrial levels and pursue efforts to reach 1.5°C (Paris Agreement, 2015), bioenergy and biofuels' share in the global energy matrix must be accelerated to achieve at least a doubling in the next ten years, even assuming much higher levels of energy efficiency, high levels of electrification of transport, and deployment of other renewables. In order to achieve the potential of bioenergy to support the bioeconomy of the future, the main requirements can be summarised as follows:

- Provide sustainable biomass supply at scales required locally and regionally.
- Identify crops that can maintain productivity on marginal land.
- Select high-yielding bioenergy crops that are more efficient in their use of water and nutrients and more resilient to climate change.
- Identify ways to integrate bioenergy production into existing activities.
- Develop market-based incentives for growing and utilising biofuels.
- Use research and development to improve conversion processes.
- Develop conversion processes of biomass to provide energy in convenient forms.

These are all issues that will be pursued in the following chapters, with a particular focus on one energy crop, *Miscanthus*, that in the last 20 years has become one of the leading contenders for the title of 'ideal energy crop'. In Chapter 2 we review the evidence for proposing that perennial rhizomatous grasses such as *Miscanthus* might be given the title of 'ideal energy crops' and in what circumstances, and more importantly where, they might be cultivated and used. In Chapter 3 we look at how the environment affects the growth and development of a *Miscanthus* crop and how we can use growth models coupled with climate and soil conditions to predict the yields of *Miscanthus* on spatial scales from field plots to continents. In Chapter 4 we will look at the issues of sustainability. One of the most important challenges to growing *Miscanthus* is to maximise its productivity, but doing so in a sustainable manner. In Chapter 5 we show how breeding efforts could produce even higher yielding crops that farmers are able to grow sustainably while having a minimal environmental footprint. In Chapter 6 we look at the range of commercial uses for *Miscanthus*, particularly as part of a developing bioeconomy. Finally, in Chapter 7 we look to the future and review how policies and markets for *Miscanthus* may develop over the rest of this century.

2 Identifying high yielding biomass crops

Introduction

Almost any form of plant material could be used as an energy crop but what is required is a plant that provides the highest possible energy yields with the minimum inputs of energy to produce that yield, that is a 'low input – high output' process. This is the 'holy grail' of bioenergy production. Of course it is not quite as simple as this because the yield needs to be in a form that is useful for energy production – can it be burnt efficiently to deliver heat or electricity or can it be converted to a liquid biofuel? The amount of energy derived from a biofuel crop depends on two factors: first the amount of dry matter of the crop that can be harvested and second the 'quality' of that dry matter in terms of the amount of energy that can be derived from it. Furthermore, the energy crop needs to be produced with as low as possible impact on the environment. This means that production and utilisation must produce the minimum of harmful greenhouse gases, while having low impact on biodiversity and not threatening the environment by acting as an invasive weed should it escape from the farmers' fields or polluting waterways by leaching phosphates and nitrates applied as fertilisers.

Depending on the origin and production technology of biomass crops they are referred to as first, second, and third generation. First generation crops are generally today's food crops from which energy-containing molecules like sugars, oils and cellulose are extracted. However they provide limited biofuel yields and their use reduces much needed and more valuable food supplies. It is now widely recognised that we need to accelerate the generation of alternative, more advanced, biomass crops as a source of bioenergy. The second generation of biofuels are now being produced from feedstocks of lignocellulosic, non-food materials that include straw, bagasse, forest residues and purpose grown energy crops. These biofuels rely on using the biomass that is not suitable to be used as food. Second generation biofuels include either plants that are specifically grown for bioenergy production, preferably on marginal lands not used for agricultural production, or inedible parts of food crops and forest trees that can be efficiently processed for bioenergy by improving current technologies. Third generation biofuels are based on algal biomass production. They are presently being extensively studied in order to improve both the metabolic production of fuels and the separation processes in bio-oil production to remove non-fuel components and to lower production costs. All three generations of biofuels are dependent on the process of photosynthesis to convert intercepted sunlight into high energy products.

What are the qualities of 'ideal' second generation biomass crops?

Plants can be viewed as a set of structures and mechanisms for 'capturing' resources from the environment. An ideal energy crop should have a sustained capacity to capture and convert the available solar energy into harvestable biomass with maximal efficiency and with minimal inputs and environmental impacts; in other words a high resource use efficiency. These characteristics are largely a consequence of the photosynthetic pathway and maximising the efficiency of light, nutrient and water use (Heaton et al., 2004b). There are three distinct forms of photosynthesis used by plants; referred to as C_3, C_4 and CAM (Crassulacean Acid Metabolism). There are far fewer species of plants with CAM type photosynthesis than the other types and they tend to be found in very dry and warm environments, like deserts. C_4 photosynthesis is the most efficient form of photosynthesis in warm to hot terrestrial environments and plants with this form of photosynthesis, such as maize, sugar cane and sorghum, have very high growth rates under warm conditions. The superior efficiency of C_4 photosynthesis is typically expressed in terms of use of the major resources that plants require, that is, light, water and nutrient use efficiency.

Ultimately, the ability to produce bioenergy crops sustainably is of paramount importance. As Somerville et al. (2010) have pointed out, because the use of groundwater for irrigation is generally not sustainable, the type of energy crop grown in a given region is primarily related to its water use efficiency (WUE). As a consequence it is likely that a relatively water-inefficient C_3 species such as poplar will be grown only where rainfall is abundant and water-efficient. PRGs such as *Miscanthus* and switchgrass will be grown where rainfall is not in excess. Arid regions are far more suitable for highly water efficient CAM plants.

Light use efficiency

The ultimate limit on biomass yield is determined by the amount of available light, its efficiency of interception by the plants, and the efficiency with which intercepted light is converted into biomass (Heaton et al., 2008a). C_4 plants have substantially reduced the energetically wasteful process of photorespiration, but at the cost of more energy being required for each molecule of CO_2 that is assimilated. As a consequence, the maximum efficiencies with which plants convert light energy, using the existing pathways of energy transduction into stored carbohydrate, are close to 6.0% and 4.6% for C_4 and C_3 plants, respectively (Heaton et al., 2008b) (Figure 2.1).

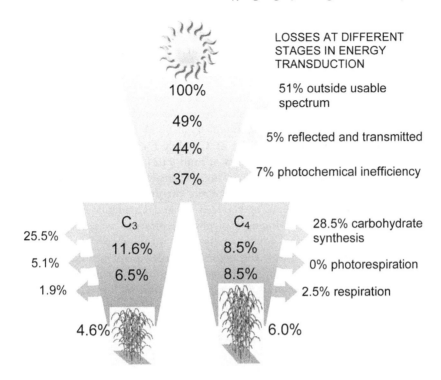

LOSSES AT DIFFERENT
STAGES IN ENERGY
TRANSDUCTION

100%   51% outside usable
         spectrum

49%

44%   5% reflected and transmitted

37%   7% photochemical inefficiency

C₃   C₄
25.5%   28.5% carbohydrate
11.6%   8.5%   synthesis

5.1%
6.5%   8.5%   0% photorespiration

1.9%   2.5% respiration

4.6%   6.0%

Figure 2.1 A theoretical analysis of the maximum efficiency of conversion of incident solar energy into biomass energy
(Adapted from Heaton et al., 2008b).

The potential yield of an energy crop can be estimated using an equation based on the principles developed by Monteith (1977): $W_h = S_t.\varepsilon_i.\varepsilon_c.\eta/k$ where W_h is the dry matter at final harvest (g m^{-2}), S_t is the incident solar radiation (MJ m^{-2}), ε_i is the efficiency with which the radiation is intercepted by the crop (dimensionless), ε_c is the efficiency with which the intercepted radiation is converted to biomass energy (dimensionless), η is the amount partitioned into the harvested components (dimensionless) and k is the energy content of the biomass (MJ g^{-1}). The quotient of dry matter yield to accumulated intercepted radiation is often referred to as the radiation use efficiency (RUE, g MJ^{-1}). While S_t is dependent on the location and k varies little between species (Beale and Long, 1995) the final dry matter produced at harvest depends primarily on ε_i and ε_c. Interception efficiency (ε_i) depends on the duration, size and architecture of the canopy. A crop that can maintain a closed canopy of viable leaves throughout the year, or at least through the period of maximum sunlight, will have the highest efficiency of interception (Monteith, 1977). In temperate regions the main factor determining this will be the ability to develop leaves rapidly at the start of the growing season (Farrell et al., 2006).

Monteith (1977) was the first to show that for healthy crops, ε_c varies relatively little within each photosynthetic group (C_3 or C_4) so that the potential dry matter productivity of a biomass crop, at a given site, will be determined primarily by the ability to form and maintain a closed canopy and by the photosynthetic type.

The difference in ε_c between photosynthetic pathways explains, in part, the superior performance of C_4 plants in warmer environments. Long et al. (2006) calculated a theoretical maximum ε_c of 0.051 in C_3 plants and 0.060 in C_4 plants at a leaf temperature of 25°C, while the maximum measured ε_c for C_3 crops over a growing season is around 0.024, and 0.034 for C_4 crops (Monteith, 1977). The highest short-term efficiencies are 0.035 for C_3 plants and 0.043 for C_4 plants, about 70% of the theoretical maxima (Beale and Long, 1995). The difference in ε_c between C_3 and C_4 plants increases with temperature because of the increase in photorespiration as a proportion of photosynthesis so that the advantage is most pronounced in the tropics (Long et al., 2006). As a result the highest recorded plant productivities are found in a C_4 perennial grass, *Echinochloa polystachia*, growing in flooded conditions in central Amazon at 100 Mg (dry matter) ha^{-1} yr^{-1} (Long, 1999). However, Long et al. (2006) using models that combine leaf photosynthesis and canopy radiation distribution show that, while the advantage of C_4 photosynthesis diminishes with temperature, there is still a C_4 advantage for canopy photosynthesis even as low as 5°C. Consequently, even in temperate climates some advantage is gained from C_4 photosynthesis and this is supported by the observation that the highest dry matter production in NW Europe has been measured for *Miscanthus* x *giganteus* that has produced as much as 29 Mg (dry matter) ha^{-1} yr^{-1} and has a measured ε_c of 0.039 (Beale et al., 1999).

Water use efficiency

To maximise productivity during the growing season, adequate supplies of water are required to maintain optimal rates of photosynthesis and to maintain green leaf area to maximise the efficiency of light interception. At the leaf level, C_4 species have, in theory, a higher WUE than C_3 species, which is largely explained by the fact that C_4 leaves have typically a 30–40% lower stomatal conductance while maintaining a higher photosynthetic rate than C_3 leaves (Long, 1999). These differences appear to be maintained at the whole plant level when WUE is expressed as a ratio of dry weight gained per unit of water transpired (Long, 1999) and it is now well established that the WUE of C_4 species is generally twice that of C_3 species, although at lower temperatures this difference is much smaller due to the reduced humidity gradient which drives transpiration. Functionally, the advantage of increased WUE is most important in conserving soil moisture and extending the period of maximum photosynthetic activity of the canopy.

Nitrogen use efficiency

Two terms are used to describe productivity per unit of nitrogen resource. While photosynthetic nitrogen use efficiency (PNUE) is the net rate of leaf CO_2 uptake in full sunlight per unit leaf nitrogen content, nitrogen use efficiency (NUE) is the ratio of increase in plant biomass to increase in plant nitrogen over the growing season. In energy crops the latter approximates to the ratio of biomass to nitrogen at the end of the growing season. In terms of field crops, NUE is determined at three levels in perennial species. First, NUE is enhanced by increasing the amount of biomass produced per unit of nitrogen invested into the photosynthetic apparatus. Second, NUE can be enhanced by increasing the fraction of soil nutrients that are assimilated by the plant (Lewandowski and Schmidt, 2006). Third, in perennial species, NUE is enhanced by increasing the fraction of nitrogen translocated out of the leaf canopy and stems during senescence; the translocated nitrogen can then be stored in the rhizomes for use in the following year. Efficient recovery of nitrogen during senescence increases the efficiency of internal recycling of nutrients. The combined effect of these properties is to both minimise the quantities of nitrogen that need to be applied as fertiliser and the amount lost to drainage water.

Because C_4 species concentrate CO_2 at the site of the carbon fixing enzyme Rubisco, the theoretical requirement for nitrogen in photosynthesis is less than in C_3 species. At the estimated concentration of CO_2 at Rubisco in C_4 plants, Long (1999) has shown that a C_4 leaf would require, at 30°C, as little as 15% of the Rubisco in a C_3 leaf to achieve the same rate of light saturated photosynthesis. The benefit of a lower requirement for Rubisco in C_4 leaves is, however, partially offset by the nitrogen requirement for the enzymes of C_4 metabolic cycle. At the whole plant level the difference in leaf nitrogen concentration between C_3 and C_4 plants combined with the higher leaf photosynthetic rate of C_4 species results in a PNUE that is approximately twice as high in C_4 compared to C_3 plants. In perennial crops this leads to a more than doubling of NUE in C_4 compared to C_3 plants.

C_4 species in cool-temperate climates

The cool-temperate climatic zone of Eurasia and North America represents a vast area that potentially could be cultivated to meet a future demand for productive bioenergy crops. C_4 species, however, are not common in the flora of the cool-temperate climate zones, indicating there may be significant yield restrictions on these plants in these regions. Physiological explanations for the relative rarity of C_4 plants in cold climates generally argue there is a fundamental restriction within the C_4 pathway that prevents photosynthetic performance at low temperature relative to what is possible in C_3 plants. These physiological restrictions include lower quantum yield, a low Rubisco capacity at low temperatures, lability of C_4 cycle enzymes, and greater likelihood of high light stress that in turn increases photo-protection costs or photoinhibition (Long, 1999). There are nevertheless, a few dozen C_4 species that are tolerant of low temperatures. These include native grasses and dicots found in high latitude and high elevation environments. In addition, varieties of the C_4 crop *Zea mays* have been successfully selected for improved cold tolerance although both the initial establishment of the crop and subsequent canopy development is still frequently limited by cool spring temperatures. Low temperatures can influence leaf photosynthesis both by reducing the efficiency of existing leaves and by affecting the development of new leaves which, as a consequence, have reduced efficiency at maturity. Typically, when leaves of C_4 plants are exposed to bright sunlight at temperatures below 15°C photoinhibition of photosystem II occurs (Long, 1983), resulting in a reduction of maximum quantum efficiency. If, on the other hand, leaves develop at low temperatures there are reduced levels of numerous thylakoid proteins and stromal enzymes which leads to a reduction in the light saturated rates of CO_2 uptake.

One group of C_4 plants which appear to be uniquely cold tolerant are the *Miscanthus* species (Jones, 2011). Unlike other C_4 species, such as sugar cane and maize, *Miscanthus* is capable of developing photosynthetically competent leaves under chilling temperatures below 10°C. Also, unlike other low temperature tolerant C_4 species it is able to achieve high efficiencies of light energy conversion and accumulate large amounts of biomass at low temperatures by maintaining physiologically active leaves at temperatures 6°C below the minimum for maize. *Miscanthus*, in contrast to almost all the other C_4 species examined, is able to realise the high photosynthetic potential of C_4 plants when grown under temperate conditions in the field in southern England. *Miscanthus* is an exception because when it is exposed to chilling temperatures there is a large increase in the content of a chemical called zeaxanthin in its leaves which protects the plant against the damaging effect of light which cannot be used in photosynthesis at low temperatures. This mechanism appears to underlie the remarkable capacity of this grass to grow in cool climates and greatly out-yield other C_4 species at these temperatures (Heaton et al., 2004b; Long and Spence, 2013). However, chilling tolerance of *Miscanthus* has been predominantly studied under controlled environment conditions when the chilling shock has been applied rapidly to leaves that have developed at non-chilling temperatures. Recently, Fonteyne et al. (2018) have shown that *Miscanthus* leaves that develop under warmer conditions and are then exposed to chilling are different metabolically from leaves developed under chilling conditions and this is reflected in their growth response to chilling. In most *Miscanthus* genotypes this leads to a trade-off between high growth rate and chilling stress tolerance. However, Fonteyne et al. (2018) have identified a remarkable response in *M. giganteus* where this genotype has overcome this trade-off with a response similar to chilling tolerant genotypes in early spring but growth rates similar to chilling sensitive genotypes later in the season.

Most of the early experimental work in characterising the chilling tolerance of photosynthesis in *Miscanthus* has been carried out on the most widely planted genotype *Miscanthus x giganteus*. More recently, Glowaka et al. (2015) asked the question whether the exceptional chilling tolerance of C_4

photosynthesis found in *Miscanthus x giganteus* can be exceeded in other *Miscanthus* genotypes. The experiment involved assessing 864 accessions from 164 different populations of *M. x giganteus, M. sacchariflorus, M. sinensis* and *M. tinctorus* collected across Japan, selected for late frost survival of their leaves in the field. Just one accession of *M. x giganteus* was identified that had a greater capacity for photosynthesis under chilling than the widely grown 'Hornum' clone. These results illustrate how rare an adaptation this cold tolerance is and that this accession will be important material for breeding new synthetic *M. x giganteus* with even higher capacity for photosynthesis under chilling.

Examples of C_4 species as biomass crops

First generation C_4 biomass crops

Currently, maize and sugarcane are the two most widely exploited examples of C_4 species used as sources of first generation biofuels, providing starch and sugar respectively (US DOE, 2016). Although grain starch from maize is currently the predominant source of biofuel in the United States, corn stover (leaves and stalks), the vegetative residue remaining after the grain is harvested, represents approximately 50% of the above-ground dry matter and could also be used as a lignocellulosic feedstock for ethanol production (de Leon and Coors, 2008). It has been estimated that approximately 256 Mt yr^{-1} of corn stover will be available in the USA by 2030 which could provide 20% of the biomass needed to replace 30% of the current transportation fuel use (Perlack et al., 2005). Future exploitation of maize as a biofuel depends on the selection of varieties with increased biomass production and bioconversion efficiency. The essential requirement for breeders is to either change the architecture of the maize plant to transform a primarily grain-producing plant to a biomass producing plant or to increase total biomass while maintaining grain yield potential. The advantage of the latter is that it would retain a high-value feed product while increasing the yield of lignocellulosic material for bioethanol production (de Leon and Coors, 2008). However, there is still a requirement to alter the cell wall composition to remove the recalcitrance to the hydrolytic enzymes required for the conversion of the polysaccharide fraction into simple sugars for fermentation. Unfortunately, attempts so far to lower the lignin content to increase the digestibility of maize stovers (i.e. leaves and stalks left in field after harvesting grain) has led to lower yields (Dhugga, 2007).

Other first generation sub-tropical and tropical C_4 biofuel feedstocks are *Sorghum bicolor* (sorghum) (Yuan et al., 2008), *Penisetum purpureum* (Napier grass) and *Erianthus spp.* (Samson et al., 2005). Sorghum can be cultivated for three processing streams: grain starch, similar to corn starch for the production of ethanol, high-sugar stem juice that can be used directly for fermentation, and dry bagasse left after juice extraction that can be used for lignocellulosic feedstock for fermentation (Yuan et al., 2008). In addition to this versatility of utilisation, sorghum is a stress tolerant species that can be grown on poor quality land with low inputs. Because industrialised nations with large biofuel targets such as the United States and European Union may not have the land needed to meet their growing demand for current first generation agricultural biofuels, there is an incentive for land-rich tropical counties to help meet these rising targets (Gibbs et al., 2008). Consequently there is an increasing opportunity for high yielding C_4 crops to be grown for biofuel feedstock in the tropics (Samson et al., 2005; Koh and Ghazoul, 2008). Napier grass has, for instance, been shown to achieve annual yields in excess of 55 Mg ha^{-1} and because of possible provision of nitrogen through biological nitrogen fixation of atmospheric nitrogen it appears that very little nitrogenous fertiliser may be necessary to maintain yields of approximately 30 Mg ha^{-1} yr^{-1} (Samson et al., 2005).

Second generation C4 biomass crops

In the United States the favoured second generation biomass crop is switchgrass (*Panicum virgatum* L.). It is a large perennial C_4 grass native to the North American prairie, that has been historically used as forage (Perlack et al., 2005; Monti, 2013). It was chosen by the US Department of Energy in 1991 as a model energy crop and in productivity trials of different varieties in a range of locations yielded an average of 13.4 Mg ha^{-1}, ranging from 9.9–23 Mg ha^{-1} (McLaughlin and Kszos, 2005).

Switchgrass was a widespread component of the tall grass prairie and occurred in non-forested areas throughout the eastern two thirds of the United Sates before the Europeans arrived (Parrish and Fike, 2005). Its original use was as forage and it is only recently, in the last 20 years, that it has been adopted as a biofuel (Parrish and Fike, 2005). The species' open pollination pattern and self-incompatibility mechanisms results in each plant in a population of switchgrass possessing a unique, heterozygous genotype. Importantly, as a native species it was considered more environmentally acceptable than intro-duced species such as *Miscanthus*. Morphologically, switchgrass is a rather course grass that grows from 0.5 to 3.0 m tall, with rooting depths of up to 3 m. The rhizomes show a good deal of variability which influences the spread of the stems to form a more bunched or open plant. There are two 'forms'; the 'upland' and 'lowland' forms which are associated with more hydric mid- to northern latitudes and drier lower latitudes, respectively. The open pollination in switchgrass has led to the development of a high level of genetic variability resulting from site-specific conditions and unique genotypes, which interact to produce a wide range of phenotypes (Casler, 2005). The possession of broad adaptation to environmental conditions, both climatic and edaphic, was a key factor in identifying switchgrass as a potential herbaceous energy crop for wide use across North America from the Atlantic coast to the Sierra Nevada Mountains (US DOE, 2006).

Most currently available commercial cultivars of switchgrass have been selected for their forage qualities, as breeders have only recently begun to select for traits that may be exploited in energy cropping (Parrish and Fike, 2005).

Field trials resulted in the selection of the variety 'Almo' as a high yielding variety which is also drought tolerant. Breeders are selecting for lines which combine broad adaptation to the environment with greater biomass productivity. Increases in biomass are most frequently linked to phenological (*i.e.* seasonal patterns of growth) and morphological traits (van Esbroeck et al., 2003). The phenological ideotype of a high yielding biomass grass is one that triggers spring growth soon after the danger of freezing injury is passed and then prolongs its vegetative activity late into the growing season but also allows time for good seed set and complete senescence before the first killing freeze in autumn.

In cool temperate regions of Europe, the sterile triploid of *Miscanthus* called *M. x giganteus* has attracted the most attention (Lewandowski et al., 2003b). Taxonomically, *Miscanthus* is closely related to several other species of high economic value such as maize, sorghum and sugarcane, in the predominantly tropical grass tribe Andropogoneae (Clifton-Brown et al., 2008). Within the Andropogoneae all species have C_4 photosynthesis of the NADP-ME type. *Miscanthus sensu lato* (*s.l.*; in a broad sense) contains approximately 14–20 species (Hodkinson et al., 1997) but its genetic limits have been re-evaluated using molecular phylogenetics (Hodkinson et al., 2002b) and has been reduced to approximately 11 species, all with a basic chromosome number of 19. In some of the earliest productivity trials, three *Miscanthus* species were identified as having the highest potential for biomass production (Jones and Walsh, 2001); these are *M. x giganteus*, *M. sacchariflorus* and *M. sinensis*. As *M. x giganteus* is a naturally occurring sterile hybrid all plantings are from the same clone. *M. x giganteus* has been wrongly called *M. sinensis* 'Giganteus', *M. giganteus, M. ogiformus* (Honda) and *M. saccariflorus var. brevibaris* (Honda). Several varieties and horticultural cultivars of *M. sacchariflorus* and *M. sinensis* have been described and they can hybridise and form a species complex with *M. x giganteus* (Hodkinson et al., 2002a; Clifton-Brown et al., 2008). This complex is considered to be the primary gene pool of *Miscanthus* available for plant breeding. The *Miscanthus* genus is native to eastern and south-eastern Asia and presumably originated in the broad area. Its natural geographic range extends from north eastern Siberia, 50°N in the temperate zone to Polynesia, 22°S in the tropical zone, and westwards to central India. It is therefore found in a wide range of climatic zones. The range of altitudinal zones are from sea level tropics where *M. floridus* is found to altitudes up to 3,100 m on dry mountain slopes in Guizhou, Sichuan and Yunnann in China where *M. paniculatus* occurs (Clifton-Brown et al., 2008).

Both switchgrass and *Miscanthus* have been shown to be high yielding crops that can be grown on low-quality, marginal land, but a quantitative review of annual production values from peer-reviewed articles describing trials of both switchgrass and *Miscanthus* in the United States and Europe found that *Miscanthus* yields on average more than twice that for switchgrass. The peak annual biomass for *Miscanthus* was 22 Mg ha^{-1} (97 observations) while switchgrass produced 10 Mg ha^{-1} (77 observations) (Heaton et al., 2004a).

Table 2.1 Characteristics of an ideal biomass energy crop present in Annuals, Short-rotation forestry and *Miscanthus*

Crop characteristics	Annuals	Short-rotation forestry	Miscanthus
C_4 photosynthesis	✓		✓
Long canopy duration		✓	✓
Perennial		✓	✓
Recycles nutrients to roots			✓
Stores carbon in soil		✓	✓
High output/input energy ratios			✓
Dry down in the field (low drying costs)			✓
High water use efficiency	✓		✓
Low mineral content at harvest (clean burning)		✓	✓
High biodiversity			✓
Non-invasive	NA	✓	✓
Few pests and diseases			✓
Good winter standing (low storage costs)		✓	✓
Use existing farm equipment for planting and harvesting	✓		✓

NA: not applicable
(based on Heaton et al., 2004b and Jones, 2011).

Is there an 'ideal' second generation biomass crop?

From the range of potential biomass crops can we identify what we might call the 'ideal' second generation energy crop? Table 2.1 lists the favorable characteristics of *Miscanthus* compared with first generation annuals such as maize and short-rotation forestry. Considering their characteristics, perennial rhizomatous grasses like *Miscanthus* can be considered as close to ideal. Their perennial nature means that once planted they have the potential to remain in the soil over winter and grow in the following spring to produce an annual crop for many years. This has several advantages; reduced soil disturbance means that greenhouse gas emissions are low, the above ground biomass can be harvested repeatedly and regrow, and nutrients are recycled to the rhizome and root system for over-winter storage when the above ground shoots senesce prior to harvest. Many of these perennial rhizomatous grasses also possess C_4 photosynthesis, which as we have seen, confers yield and resource efficiency advantages.

The focus of this book is *Miscanthus*, a small group or genus of perennial rhizomatous grasses. The reason why the focus is on *Miscanthus* is because I would argue that these grasses are the closest we are to finding 'the ideal' second generation energy crop. The advantages of *Miscanthus* can be summarised as follows, (1) it produces large amounts of biomass in a wide range of climates from cool temperate to sub-tropical, (2) it has a chemical composition that suits it for a range of conversion processes, (3) it has a positive environmental footprint, including low fertiliser requirement and low greenhouse gas emissions (particularly N_2O) and (4) it is largely non-invasive.

The history of *Miscanthus* as a biomass crop

Miscanthus was first proposed as an energy crop by the horticulturalists who planted it as an ornamental grass and commented on its very high growth rates in the gardening literature. During the 1970s researchers in bioenergy production began to take an interest in *Miscanthus* and they established a large number of productivity trials with the aim of assessing the yield potential of a number of candidate species including *Miscanthus*, particularly in Denmark and Germany. These trials mainly used one particular cultivar of *Miscanthus* which was subsequently identified as a hybrid between *Miscanthus sinensis* and *M. sacchariflorus* (Hodkinson et al., 1997), and was called *M. x giganteus*. The crop was either harvested at the end of the growing season to obtain maximum biomass yields or allowed to dry down in the field for a delayed harvest. This saves on drying and storage costs but reduces harvestable yield due to leaf loss following senescence and weathering. The trials grew *Miscanthus* over a range of climatic conditions and soil types and also added a range of amounts and types of fertilisers. Most of these trials confirmed the surprisingly high harvestable yield of *Miscanthus* under most conditions and the remarkably low requirement for fertilisers to achieve these yields. In order to factor out the effect of harvest date the delayed yields were scaled to September 1st values by calculating a rate of senescence. On average *M. x giganteus* yields decreased by 0.01 Mg ha^{-1} day^{-1} for each day harvest was delayed after September 1st (Price et al., 2004; Lewandowski et al., 2003b). In practice, *M. x giganteus* is normally harvested late in winter or in early spring so losses may be considerable.

The yield potential of *Miscanthus* in Europe and the USA

Although yield trials were underway from the 1970s in Denmark and Germany it was not until the 1990s that a more systematic approach was adopted and in particular the European *Miscanthus* Network brought together 16 European partners to establish field trials across Europe (Lewandowski et al., 2000; Jones and Walsh, 2001). Results of these and additional trials indicate harvestable *Miscanthus* yields range from 10–25 Mg ha^{-1} yr^{-1} in northern Europe and from 25–40 Mg ha^{-1} yr^{-1} in central and southern Europe, although here irrigation was frequently required (Lewandowski et al., 2000; Clifton-Brown et al., 2001; Price et al., 2004; Christian et al., 2008). Other significant conclusions from these trials were that (i) winter survival of newly established *M. x giganteus* was unreliable in much of northern Europe and that this was most likely associated with the slow development at low temperatures of rhizomes in the first growing season, (ii) moisture content of autumn-harvested *Miscanthus* is generally higher than in spring but harvestable yields in spring are 30–50% lower than in autumn so that optimum harvest time depends on the climatic conditions at the site, (iii) nitrogen fertiliser applications generally produced only a small yield response, probably as a result of efficient nitrogen cycling in and out of the rhizomes and the lower requirement of plants with C_4 photosynthesis for nitrogen, and (iv) higher planting density increases the rate at which a ceiling yield is reached, but cooler northern sites still require three to four years, while southern sites normally reach a ceiling in two years.

In the USA, current knowledge of the yield potential for *Miscanthus* is primarily derived from a relatively small number of small-scale and short-term studies. The first synthesis of plot trial yields of *Miscanthus* was reported by Heaton et al. (2004a), who used data from 21 trials and found that the annual average yield was 22 Mg ha^{-1}. To assess how *Miscanthus* performs across a wider variety of locations and for a greater number of years in the USA, the US Department of Energy/Sun Grant Initiative Regional Feedstock Partnership was initiated in 2008. The objective was to develop national maps of potential yields for a number of herbaceous species including *Miscanthus* (Lee et al., 2018). The intention was to use field-scale plots and traditional agricultural machinery because smaller scale studies often overestimate yield potential; however because insufficient rhizomes for vegetative propagation were available for planting at field scale, small 10 x10 m plots were used. The results of these trials indicated that yields of *Miscanthus* can be sustained at or above 15 Mg ha^{-1} across most years, locations and fertiliser treatments, and that certain favorable conditions allow this to be exceeded to produce yields greater than 25 Mg ha^{-1}.

However, concerns about yields from small plot trials overestimating field scale yields remain. For example, Searle and Malins (2014) reviewed the yield expectations of commercial planting of energy crops from published material in Europe and the USA and pointed out that yields at the field scale are frequently significantly lower than small plot measurements. The reasons for this are complex but may be associated with edge effects, where plants at the edges of small plots intercept more light and grow faster than plants in the centre of larger plots, as well as the crops being planted on good quality rather than marginal land. However, probably more significant are the large senescence losses before harvest, and losses post-harvest, which mean that dry matter delivered for processing is very much lower than the peak yields that are frequently quoted from the type of plot trials described above. Searle and Malins (2014) therefore concluded that yield expectations at commercial scale on marginal land are much lower than those extrapolated from the plot trials. They suggest that expected yields may be as low as between 3 and 5 Mg ha^{-1} in cool temperate regions and 7–15 Mg ha^{-1} in temperate regions.

Conclusions

There is now reliable evidence from a very large number of field trials that *Miscanthus* can produce higher harvestable yields than any other putative bioenergy crop in a wide range of environments. Given the evidence presented in this chapter, I would argue that *Miscanthus* is indeed an 'ideal' second generation biomass crop. However, whether *Miscanthus* is the crop of choice at any location is not determined solely by its yield. Clearly it is an extremely important consideration but not the only factor. Choices about production of biofuels, while strongly influenced by maximising harvestable yield of the bioenergy crop, will be determined by a whole range of other factors which influence the production chain from field-to-wheel. We will consider these in later chapters. However, in the next chapter we look in more detail at some of the factors that influence the production of *Miscanthus* and review how this information is used in growth models to predict the landscape scale production potential of *Miscanthus* and its future contribution to biomass energy production on a global scale.

3   Controls of *Miscanthus* productivity

Introduction

Plants have evolved and adapted to remain productive in a wide range of environments, and the research by environmental physiologists and agronomists has greatly advanced our understanding of the physiological processes that ultimately determine yield in any particular environment. This research has also provided the basis for scaling the physiological processes from the plant tissue to the canopy, ecosystem, region and globally. In this chapter we review what we know about the influence of environmental conditions on the growth and development of *Miscanthus* and how these combine and interact to determine yield of the crop. We focus on identifying the particular characteristics of plants like *Miscanthus*, which lead us to describe them as the 'ideal' biomass crops.

The origins of the genus *Miscanthus* in the tropical and subtropical parts of south-eastern Asia, which is characterised by warm temperatures and heavy and well distributed rainfall, suggests that the natural climatic preference of *Miscanthus* would be mild temperatures and high water availability. Throughout these regions *Miscanthus* is generally considered to be a weed of disturbed areas, but it also forms extensive grassland communities. The *Miscanthus* hybrid, M. *x giganteus*, which is now the most widely cultivated cultivar of *Miscanthus* was first collected in Yokohoma, Japan in 1935 from where it was taken into cultivation in Denmark by Karl Foerster. This plant was subsequently distributed throughout much of cool temperate Europe where it was found to be remarkably cold tolerant. In fact unlike the vast majority of closely related species with a similar C_4 photosynthetic pathway M. *x giganteus* is capable of developing photosynthetically competent leaves under chilling temperatures well below 10°C. Also, unlike some other low-temperature tolerant C_4 species it is able to achieve high efficiencies of light energy conversion and accumulate large amounts of biomass at low temperatures by maintaining physiologically active leaves down to almost freezing.

The growing cycle of *Miscanthus*

Most agronomic and physiological research on *Miscanthus*, and in particular productivity trials, have focused on clonal material of M. *x giganteus*, but the two species of *M. sinensis* and *M. sacchariflorus* have also featured in a number of studies. Currently the main method of propagation of *Miscanthus x giganteus* is as rhizomes or rhizome-derived plants (Xue et al., 2015). Rhizomes are taken from established plants and for rhizome-derived plants are grown on in greenhouses during the winter and spring. This is an expensive method for establishing *Miscanthus* and in some circumstances results in large plant losses during the first growing season due, in particular, to frost damage of rhizomes in their first winter. Furthermore, the multiplication ratio is very low. Other plant parts can be used for propagation, including nodal buds and nodal stem cuttings, but these methods are slow and relatively expensive given their low multiplication ratio. Micropropagation has also been used successfully but again this is a relatively expensive operation. More recently there has been a concerted effort to propagate from vegetative plant tissue using a process called CEED (Crop, Expansion, Encapsulation, and Delivery system) developed by New Energy Farms (http://newenergyfarms.com/products/ceeds). CEEDS^{TM} is an 'artificial seed' system in which plant tissue is encapsulated in a growing medium which can contain growth promotors and crop protection products. The artificial seeds can be planted using conventional equipment or by using equipment developed by New Energy Farms to automate planting, even into minimum-till and no-till seedbeds. It is claimed that, among other advantages, establishment costs are substantially reduced and that plants are more vigorous after establishment.

Establishment of the crop in field trials is normally in spring or early summer (March to May in the northern hemisphere). Because of the high production costs of establishing from rhizomes or plantlets the optimum planting density is low, normally 2–5 plants m^{-2} (Caslin et al., 2011; Zub and Branscourt-Hulmel, 2010). After emergence the number of shoots per plant increases rapidly in May, June and July up to 40 stems m^{-2} (Bullard et al., 1997), but during the growing season the number of productive shoots falls to about 25 stems m^{-2}. These stems continue to grow until the autumn when they begin to senesce and growth normally stops with the first frost.

During the growing season the leaf area development increases exponentially to reach a maximum leaf area index (LAI, leaf area per unit ground area) of 6–10, depending on the climate. A mature stand of *M. x giganteus* is able to intercept around 90% of incident radiation when the LAI reaches 3.2 (Clifton-Brown et al., 2002). At the end of the growing season, nutrients and photosynthates are remobilised from the leaves and stems to the rhizomes. The growing cycle of a *Miscanthus* stand is strongly influenced by temperature. Growth in the spring starts when a threshold temperature is reached and the developmental cycle is regulated by thermal time, which is the cumulative value of degree days from emergence to senescence (Clifton-Brown and Jones, 1997). Figure 3.1(i) shows the canopies of four genotypes of *Miscanthus* photographed on the same day close to the end of the growing season and Figure 3.1(ii) shows the progression of leaf area index through the growing season for the same four genotypes. The annual peak yields occur in the autumn and if left unharvested decline through the winter due to leaf and stem loss. Delaying harvesting until the spring results in yields that are 25% to 50% lower than in the

Figure 3.1(i) Photographs showing the different canopy architectures of four genotypes of *Miscanthus:* (a) Gig-311, (b) Sac-5, (c) Sin-11 and (d) Goliath taken on the same day in September in Aberystwyth, Wales. Sin-11 and Goliath are genotypes of *M. sinensis*, Sac-5 is a genotype of *M. sacchariflorus* and Gig-311 is a genotype of *M.x giganteus*
(From Davy et al., 2017).

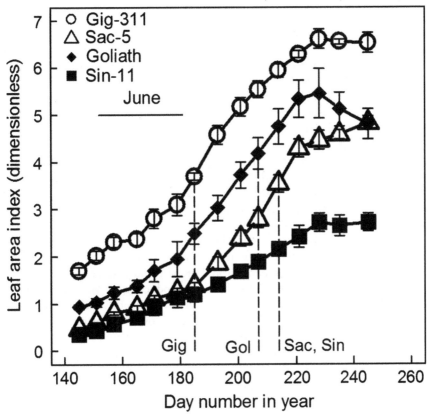

Figure 3.1(ii) The seasonal development of the leaf area index (LAI) of the canopies of the four genotypes of *Miscanthus*. The error bars are ± one standard error of the means. The vertical lines show when the canopies intercepted 90% or more of the incident radiation. The horizontal bar indicates June, when the peak radiation levels occurred
(From Davy et al., 2017)

autumn. However the advantage of delayed harvest is that moisture content declines significantly and the nutrient content declines due to remobilisation to the rhizomes.

Once established, it is expected that the *Miscanthus* crop will continue production for at least 20–25 years. During the first one or two growing seasons the yields are low but this is followed by several years when yields peak and are relatively stable although they are dependent on the prevailing weather conditions (Jones et al., 2016). In the early years of field trials there was scant information on the long-term sustainability of the *Miscanthus* crop but recently Lesur et al. (2013) carried out an analysis of long-term yield trends of *M. x giganteus* grown in Europe, drawing on data from 37 field trials at 16 sites spanning five countries. Statistical models were used to describe yield evolution in each of the trials. The models were used to estimate three characteristics of yield trends over year *viz.*

(1) the maximum yield reached across years, (2) the growing season when the maximum yield was reached, and (3) the yield decrease associated with the period of decline. The duration of the establishment phase to maximum yield was variable and shown to be strongly determined by the planting method (Lesur et al., 2013). Maximum yields were found to be highly variable and this variability was largely explained by climatic factors, demonstrated by a strong relationship between maximum yield and latitude. Model fits to the yield data showed that yield evolution was best described by a decline over time but the decline intensity was variable so that in some trials yields remained nearly steady for up to 20 years while in others decline was severe after 5–10 years. In the mid-western USA, Arundale et al. (2014a) have also investigated the yield decline in *Miscanthus*. Using data from several locations across Illinois over a period of 8–10 years they found that yield initially increased until it reached a maximum during the fifth growing season and then declined to a stable but lower level in the eighth season. The possible causes of the decline were ascribed to exhaustion of the soil nutrients, especially nitrogen, soil compaction and/or pest and disease pressure.

To summarise, the above-ground biomass production of *Miscanthus* in any year is dependent on the duration of the growth period. After the first year following planting, the start of the growing season in temperate climates is determined by the time of the last spring frost and the end of the growing season is determined by flowering time or the first winter frost, depending on the harvest date or location. The growing season cycle ranges from six to nine months, depending on the genotype. For example, *M. sinensis* genotypes generally flower earlier and therefore have a shorter growing season than hybrids of *M. sinensis* which flower later, whereas *M. x giganteus* is unable to flower under Northern European conditions so that it continues to grow until exposed to the first frost (Clifton-Brown et al., 2001). There is clear evidence for a long-term decline in annual yields but the timing of this is quite variable. In some locations, yields may remain nearly steady for up to 20 years or more whereas in other locations a severe yield decline has occurred by this time (Lesur et al., 2013).

Abiotic controls of growth and yield

The main environmental controllers of *Miscanthus* productivity are temperature, soil nutrient supply and water availability. Their impacts on growth and yield are reviewed independently in the following sections although it is important to appreciate that at any given time in the growing season growth is regulated by interactions between these controllers.

Temperature

Although *Miscanthus* is better adapted than almost all other C_4 plants to low temperatures, many trials, particularly in Europe, have shown that yields were significantly lower in cooler northern Europe than in southern Europe, most significantly when the crops in southern Europe were irrigated to avoid water stress. Growth of most *Miscanthus* species and genotypes is initiated when ambient temperature rises above a threshold of 5–10°C; this is when the leaf canopy begins to grow. Subsequently, the rate of canopy development as a result of leaf growth is controlled by temperature (Farrell et al., 2006). Experiments investigating the thermal response of leaf growth of a number of genotypes have shown that the response of absolute extension rate to temperature varies widely between genotypes. The rates of canopy photosynthesis are also temperature dependent and unlike almost all other C_4 species, *Miscanthus* is capable of developing photosynthetically competent leaves at as low as 10°C. It is also able to achieve high efficiencies of light energy conversion and accumulate large amounts of biomass at much lower temperatures than other C_4 crops such as maize, sugarcane and sorghum (Naidu et al., 2003).

As temperature regulates the length of the growing season and the rate of canopy development of *Miscanthus*, in broad terms, it determines the dry matter production of the crop. This correlation appears to be strongest in the temperate climates of northern Europe such as the UK but it breaks down in continental climates where high net photosynthetic rates and high water use efficiency are the most important traits influencing yield (Nunn et al., 2017). As we see later, these differences in traits, critical for influencing yield, are important considerations in selecting genotypes best suited for growth in different climatic regions.

Freezing temperatures are important regulators of *Miscanthus* growth and development and a capacity for winter survival has been described as the principle obstacle to the establishment of *Miscanthus* crops, particularly in temperate climates (Christian and Haase, 2001). *Miscanthus* crops encounter two types of freezing conditions that influence establishment; the long-term freezing of soils during the winter and late air-frosts during the spring. The occurrence of freezing temperatures in winter strongly influences the winter survival of the *Miscanthus* crop. Young plants during their establishment appear to be very vulnerable and in northern regions, winter survival in the year after planting appears to depend upon the sequestering of a critical amount of metabolic reserves in the rhizomes at the end of the previous growing season. It is also the case that the ability of *Miscanthus* to survive extended periods of frost is a result of the development of dormancy. A sequence of processes such as the translocation of reserves to the rhizomes, the senescence and death of the stems and leaves, combined with the increasing dehydration of the rhizomes, leads to dormancy. Dormancy is induced at the end of the growing season by the increasingly colder temperatures and longer night lengths of autumn and early winter (Zub and Brancourt-Hulmel, 2010). In the case of late air-frosts in spring and early summer, these freezing temperatures can damage and kill newly expanded leaves during the first year of the crop or the regrowth of leaves on older plants.

Soil nutrients

A characteristic of perennial herbaceous rhizomatous grasses like *Miscanthus* is their ability to continuously re-mobilise nutrients between various organs of the plant as the growing season progresses. Although the life span of the plant can be more than 20 years, its stems and leaves function for only one season. The only permanent organ is the rhizome which functions in vegetative propagation and the storage of nutrients. The internal recycling of nutrients between above- and below-ground organs allows the harvesting of biomass with a low nutrient content, but also reduces the demand for nutrients for renewed growth which is normally met through application of fertilisers.

The nutrient normally applied in greatest abundance to crops is nitrogen but another feature reducing the demand for nitrogen in *Miscanthus* is the C_4 photosynthetic mechanism which relies on a very effective photosynthetic enzyme, PEP-carboxylase for assimilating CO_2 (Jones, 2011). The presence of PEP-carboxylase means that for the same amount of photosynthesis as non-C_4 plants, *Miscanthus* has to allocate far less nitrogen to producing the enzyme required for CO_2 fixation.

As a result of it utilising C_4 photosynthesis and achieving an efficient re-cycling of nutrients, *Miscanthus* has very low nitrogen requirements and limited need for fertiliser applications. Indeed, in some agronomic trials there is evidence that *Miscanthus* can be cropped for several years before there is a requirement for added nitrogen fertiliser. For example, Christian et al. (2008) grew *M. x giganteus* for 14 successive harvests in the south of England and found that fertiliser nitrogen (N) application had little influence on yields, which reached 17.7 Mg ha^{-1} at their peak. It was suggested, however, that other nutrients such as phosphorus and potassium should be added at low levels to avoid depletion of soil reserves. In contrast to this evidence for little or no requirement for nitrogen, other trials, particularly in the USA, but also in Mediterranean Europe (Cosentino et al., 2007), have shown a significant increase in yields of stands of *Miscanthus* when nitrogen fertiliser was added. For example, in a trial in Illinois USA, Arundale et al. (2014b) found that *M. x giganteus* yield increased significantly from 23.4 Mg ha^{-1} with zero fertiliser to 28.9 Mg ha^{-1} (+25%) at an annual application rate of 202 kg N ha^{-1}. However the proportional increase in yield per unit of added nitrogen is small in *Miscanthus* compared with other C_4 crops such as maize and switchgrass (Heaton et al., 2004a), and consequently it is suggested that this response to added fertiliser is unlikely to be economically worthwhile. In conclusion, it appears that there are requirements for low levels of fertiliser to maintain yields of *Miscanthus* but there are very large site to site variations which make generalisations and advice on optimum fertiliser rates extremely difficult to make.

An important recent observation from field trials is that a non-artificial fertiliser contribution to the nitrogen supply to *Miscanthus* in the field is the result of nitrogen 'fixing' associations known as diazotrophs. Diazotrophs are bacteria and archaea that convert atmospheric nitrogen to ammonia, making it available to other life forms. Diazotroph communities have been shown to be associated with the rhizosphere and endophytic communities of *Miscanthus*, similar to many other grasses (Keymar and Kent, 2014). In a trial in the USA they were shown to provide 16% of the nitrogen by biological nitrogen fixation in the first year growth of *M. x giganteus* and it has been suggested that soil amendments, such as bio-available iron that stimulate the production of diazotrophs, could improve the sustainability of *Miscanthus* production (Soman et al., 2018). Further evidence comes from modelling the nitrogen budget of *Miscanthus* using an ecosystem process model, where Davis et al. (2010) showed that a previously unknown source of nitrogen was necessary to balance the plant nutrient budget in the crop. They therefore concluded that significant nitrogen fixation contributed to the nitrogen demand of *Miscanthus*.

Water

As a consequence of focussing on maximising yields it is inevitable that increased productivity will result in higher water demand and water supply may become a limiting factor for crop productivity. However plants like *Miscanthus* that have C_4 photosynthesis have a higher water use efficiency than plants with C_3 photosynthesis. As a consequence they produce approximately twice as much biomass for the same amount of water use (Jones, 2011). Functionally, the advantage of a higher water use efficiency is that it conserves soil moisture and extends the period of photosynthetic activity of the canopy for longer before photosynthesis is reduced by a developing water stress. Despite higher water use efficiency it is possible that growing *Miscanthus* will use more soil water reserves than the vegetation that it replaces, mainly as a result of a longer growing season and higher productivity. When Vanloocke et al. (2010) modelled the hydrologic cycle of the Midwest USA converted to *Miscanthus* production their simulations showed that on an annual basis *Miscanthus* uses more water than the ecosystem that it will replace but that the actual timing and magnitude of increased water loss to the atmosphere depends on location. The results show that as the biofuel industry transitions to highly productive cellulosic feedstocks, it will need to account for the role of water as a potential limiting factor in sustainable biofuel production.

Although early trials with *Miscanthus* showed that it could be successfully grown without irrigation in northern and middle European regions, irrigation has been found to be necessary for higher growth rates and biomass yields in southern Europe. Zub and Brancourt-Hulmel (2010) demonstrated that the supply of irrigation had a much more beneficial effect on yields in southern Europe (France and Sicily) than in northern Europe (UK) and that the effect was higher with an autumn harvest than a winter harvest. In general, under southern European conditions *Miscanthus* starts growth from the middle to end of April. Because of the warm conditions, initial plant growth is rapid up to the end of June and no irrigation is required as plants draw on soil reserves. However, irrigation stimulates further growth when applied from late June to the end of the growing season. Although *Miscanthus* yields are dependent on irrigation rates, it has been shown that doubling irrigation rates may increase yield by as little as 10%. It is therefore considered more economically and environmentally sensible to apply only moderate rates of irrigation and forgo the relatively small advantage of expensive irrigation (Lewandowski et al., 2000).

In a review of several trials throughout Europe Heaton et al. (2004b) demonstrated a significant effect of water availability on *Miscanthus* biomass production. Under varying levels of nitrogen inputs (between 60 and 240 kg N ha^{-1}) biomass production increased by between 25% and 84% with irrigation (Zub and Brancourt-Hulmel, 2010). In the UK, Price et al. (2004) measured the yields of well-established *M. giganteus* growing at seven widely dispersed sites and found significant interannual differences that were largely attributable to water stress. They developed a simple predictive model of *Miscanthus* yields and derived maps of potential and water limited yields for England and Wales. The average modelled estimates of dry matter yield at harvest on arable land in England and Wales were in the range of 6.9–24.1 Mg ha^{-1} year^{-1}. As a result of this sensitivity of *Miscanthus* to water stress, the identification of drought-tolerant genotypes that can produce more biomass under water stress conditions remains an essential requirement in any breeding programmes (Clifton-Brown and Lewandowski, 2000).

Biotic controls of growth and development

Biotic refers to anything that is living and the presence of living organisms in the environment of a plant can have both negative and positive effects. Pests and diseases generally have negative effects but the presence of certain soil bacteria, particularly endophytic bacteria, can have significant positive effects.

Pests and diseases of Miscanthus

Miscanthus species are very robust and suffer from few pests or pathogens. In Europe and the USA, hardly any cases of significant harvest losses by biotic agents have been documented. Many microorganisms growing on *Miscanthus* have been isolated during post-harvest storage of *Miscanthus* leaves (Shrestha et al., 2015). However, the main aim of these studies was to investigate their potential as a source for enzymes that can contribute to biomass conversion. Other studies describe isolated cases of pathogenesis. For example, in the 1990s, *Miscanthus* leaf blight caused by the fungus *Stagonospora miscanthi* was found in commercial nurseries and residential ornamental landscapes in Maryland, USA. And in Europe, on Belgian land used for biomass production, several *Fusarium* species, including *F. avenaceum*, *F. culmorum* and *F. miscanthi* were isolated from fields showing poor emergence of *Miscanthus* (Scauflaire et al., 2013). In pathogenicity tests under greenhouse conditions, isolates displayed the ability to cause rhizome rot disease. Another pathogen, causing leaf rust, was isolated and characterized in Korea, where the causal agent was identified as the basidiomycete fungus *Puccinia miscanthi* (Kim, 2015). In general, most of the isolated microbial pathogens are closely related to those found on sugarcane and/or other grass family crops, and it has been observed that pathogens isolated from sugarcane are, within *in vitro* tests, pathogens to *Miscanthus*. Also, invertebrate pests have been isolated from *Miscanthus*, including nematodes and aphids (Bradshaw et al., 2010). However, it is not clear how far these organisms contribute to reductions in crop yield, if at all. There is a danger that, in time, when *Miscanthus* cultivation is scaled up, pathogens and pests from other crops may start to cause yield losses and therefore pose a threat to the economic viability of the crop.

Endophytic bacteria

Endophytic bacteria live within plant tissue gaining nutrients and/or shelter without causing visible detriment to the host and are known to confer a range of benefits to the plant host, including increased growth rate, biological nitrogen fixation as referred to above, and soil phosphate solubilisation. They have been demonstrated to enter the plant through spaces between root cells but an alternative route to pass from one generation to the next is through seeds. Endophytic fungi are also known to defend against pathogens in plants by several mechanisms. One way by which they contribute to plant defence is by triggering host resistance, inducing expression of pathogen immunity genes. Consequently, endophytes can be used as biocontrol agents. In such systems, the endophyte is inoculated/introduced into the plant and once established it either prevents the pathogen from entering the plant or limits its pathogenicity once the plant is infected (Busby et al., 2016).

Evidence from productivity trials

The primary purpose of productivity trials is to assess the technical potential that *Miscanthus* has as a source of biomass for energy at the farm and landscape scale. Productivity trials provide estimates of the potential for local and regional supplies of biomass across a variety of site conditions and they also improve our understanding of how soils, climate, genetics and crop management practices, like additions of fertiliser and irrigation, influence biomass production.

A number of productivity trials of *Miscanthus*, mainly focusing on *M. x giganteus,* have been carried out in the USA and across much of Europe. However, because *Miscanthus* is relatively new as a crop, the trials have been relatively limited in geographical scope, considering the potential range over which it could be grown. Furthermore, as *Miscanthus* is grown as a long-lived perennial, many of the trials have been of too short a duration to provide reliable information on sustainability of yields. There have been a number of publications reviewing and summarising the results of these trials. Notably, Heaton et al. (2004a) carried out a quantitative review comparing the yields from a large number of trials carried out in the 1990s of the two candidate C_4 perennial crops, *M. giganteus* in Europe and switchgrass in the USA. The review was aimed at comparing yields of the two candidate energy crops and assessing the main agronomic influences on yields. Both crops showed a significant positive response to water (rainfall) and nitrogen levels but not to temperature. However *M. giganteus* yielded significantly more harvestable biomass than switchgrass across the range of all three climatic regions. Overall, *M. giganteus* was found to yield more than twice that of switchgrass (22.4 +/- 4.1 versus 10.3 +/- 0.7 Mg ha^{-1}). While both crops responded to increased precipitation, the yield of *M. x giganteus* was more strongly affected. Conversely, the switchgrass showed a stronger response than *M. giganteus* to nitrogenous fertiliser applications.

Lewandowski et al. (2003b) also reported on yield trials of *Miscanthus* across Europe at five sites, from Sweden in the north to Portugal in the south. Yields were determined in the autumn of the third year after planting, when the stands had reached maturity. Delayed harvests were also made in late winter/early spring. Autumn mean yields varied from 12.1 Mg ha^{-1} in England to 26.5 Mg ha^{-1} in Portugal and a delayed harvest resulted in a mean yield reduction of ~35%. However, late harvesting had the advantage of reducing moisture content by 48%, ash by 37%, potassium by 55%, chloride by 75% and nitrogen by 20% (Lewandowski et al., 2003a).

Because of the widely recognised need to grow energy crops on marginal land it is unfortunate that few productivity trials have been carried out on eroded and saline soils. However, Yost et al. (2017) reported yields of *M. giganteus* on eroded (claypan) soils in Missouri, USA which are marginal for grain crops. They found that yields varied from 13.3 Mg ha^{-1} on unfertilised land to 23.8 Mg ha^{-1} on fertilised land, the latter being remarkably close to yields reported on high quality land by Heaton et al. (2004a). They suggested that the level of nitrogen fertiliser application required to maximise yield was relatively low at 67 kg N ha^{-1} y^{-1}.

In other productivity trials for *M. giganteus*, carried out in Midwestern USA by Lee et al. (2017), they found that an application of 60 kg N ha^{-1} produced average yields over a four year period of 22 Mg ha^{-1} compared to 11.8 Mg ha^{-1} for unfertilised plots. The conclusion was that at this site nitrogen fertilisation was necessary for sustainable biomass production, but an application as low as 60 kg ha^{-1}N was adequate for maximum biomass yields. However, the results also suggest that site-specific nitrogen management might be necessary for the sustainable production of *Miscanthus* and that in some cases the interaction with water stress is important as nitrogen may reduce the impact of reduced soil water content by stimulating root growth and increasing water extraction from depth.

Productivity trials have also investigated the effects of management on yields of *Miscanthus*, and in particular the effects of time of harvest. These have been important in elucidating the impacts of delayed harvest on yield and quality. For example, Lewandowski and Heinz (2003) found that with a delayed harvest, bioenergy yields decreased by 14–15% between December and February and by a further 13% between February and March but this was accompanied by a significant decrease in moisture content. The delayed harvest clearly improved the combustion quality of the biomass but it also led to lower energy yield because of reduced biomass and illustrates the need for the biomass producer to optimise harvest date to achieve a balance between quality and yield. In other trials, O'Flynn et al. (2018) have highlighted a lack of understanding of the damaging effects of traffic during harvest on subsequent *Miscanthus* shoot emergence. Soil compaction is often a stress factor that negatively effects plant growth and reduces productivity. In a field trial O'Flynn et al. (2018) found that yield losses which occur after early harvest, when compaction is greatest, are substantially lower than those which occur following a late harvest in spring when shoot emergence has already started, as the magnitude of late harvest shoot damage exceeds the effects of other potentially beneficial factors. They conclude that the greatest factor in the minimisation of reduction in growth and yield after *Miscanthus* harvests is the avoidance of late harvest shoot damage.

Modelling *Miscanthus* yields

As we have seen, compared with long established crops, productivity trials of *Miscanthus* are limited in number and extent. In introducing dedicated energy crops it is important to be able to forecast their productivity and stability under a wide range of different environments including changing climate, with a particular emphasis on marginal land. There is therefore a need to develop and parameterise crop models that can provide reliable predictions of carbon assimilation, growth and yield of *Miscanthus* (Robertson et al. 2015). In this modelling process the productivity trials still play an important role in validating the outputs from the models.

Nair et al. (2012) carried out a survey of the literature and identified 14 models that have been used to simulate bioenergy crops. All of them simulate biomass production, but only a small number simulate soil water, nutrient, and carbon cycle dynamics and could be classified as ecosystem models. Broadly speaking, models have been developed using either an empirical or mechanistic approach or a hybrid of the two.

Empirical models

An empirical approach involves statistical extrapolation of plot measurements to larger areas. For example Richter et al. (2008), developed an empirical yield model based on a multiple regression of experimental yields from 14 experiments across 10 different sites in the UK, with soil available water capacity (SAWC) and long-term monthly climate data (precipitation, temperature, radiation) for each field. Harvestable yields of *Miscanthus* established for at least three years across the UK ranged from 5 to 18 Mg ha^{-1} and the overall national average yield was 9.6 Mg ha^{-1}. They concluded that in the UK, *Miscanthus* yields are clearly water-limited in many areas and that spatial and temporal variation in yield can be explained by water availability. Richter et al. (2016) combined empirical yield modelling with soil mapping and remote sensing (satellite imagery) to assess on-farm productivity of *Miscanthus* and used on-farm yield surveys at the field and farm scales to verify outputs of the productivity models which had been parameterised and validated using experimental plot-scale observations. This work identified a 'yield gap' between the actual achievable on-farm yields and the model predictions. The actual on-farm yields averaged around 9.0 Mg ha^{-1}, with a coefficient of variation of 34%, while the yield potential estimated using the empirical model averaged 11 Mg ha^{-1}. The yield-gap was larger on clay soils than on sandy or loamy soils (37% v 10%). It was concluded that overall in the UK, heavy soils are potentially the most productive with yields greater than 16 Mg ha^{-1} but they also had the most variable stand density (patchiness) which is most likely influenced by wet and dry soil at planting. This patchiness was a cause of reduced yields.

Hybrid models

Many models actually cross the boundaries between empirical and mechanistic models. A modelling and mapping system that falls into this category is called PRISM-ELM and has been developed by Daly et al. (2018). PRISM-ELM is a hybrid model that uses both empirical and mechanistic approaches to determine how climate and soil characteristics affect the spatial distribution and long-term production patterns of potential biomass resources across the coterminous states of the USA. However, unlike mechanistic models this approach does not require detailed data on plant characteristics and physiology. Daly et al. (2018) suggest that this form of hybrid model is becoming more powerful as high quality climate, remote sensing, land-use and soils data become available. The hybrid model employs a limiting factor approach, where productivity is determined by the most limiting of the factors addressed in sub-models that simulate water balance, winter low-temperature response, summer high-temperature response, and soil pH, salinity, and drainage. Yield maps are developed through linear regressions relating soil and climate attributes to reported yield data (see Figure 3.2). The conclusion from this modelling exercise was that earlier yield projection maps of Miguez et al. (2012) using a mechanistic model could be revised to show the greater regional suitability of *Miscanthus*. The resulting maps are used as inputs to analyses comparing the viability of a range of biomass crops, including *Miscanthus*, under various economic scenarios across the United States. In general PRISM-ELM maps show regional patterns that are more similar to those produced by mechanistic models than with statistical models, probably as a result of more effective treatment of water balance and effects of water stress (Daly et al., 2018). The areas of greatest disagreement among models are located along the edges of the crop's distribution, where validating field trials are largely absent.

Mechanistic models

In mechanistic models, predictions of yields are developed from first principles of plant processes including photosynthesis, respiration and assimilate distribution. The theoretical potential yield of a crop is controlled only by the biophysical limits in the location it is cultivated, so that the model is driven by environmental and edaphic parameters. At larger scales ecosystem process-based models have

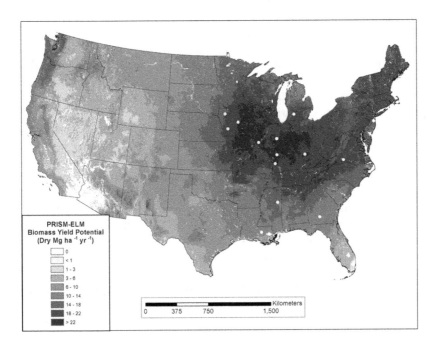

Figure 3.2 Miscanthus predicted yields, mapped across the USA using the PRISM–ELM
    hybrid model
(From Daly et al., 2018).

been developed to simulate ecosystem carbon, nitrogen and water dynamics
through the descriptions of physiochemical processes. Such models can be used to
extrapolate crop performances from farm scale to regional scales to assess global
scale production potential and the environmental impacts of energy systems.
Generally, these models have similar process descriptions but the parameters in
the models can be adjusted to match the site specific data. Table 3.1 lists some of
the published *Miscanthus* mechanistic models and their basic characteristics.

 An early attempt at mechanistic modelling of *Miscanthus* was MISCANMOD,
developed by Clifton-Brown et al. (2000). This was a simple spreadsheet-based
model, using the robust and straightforward approach of assuming that there
is a direct link between intercepted radiation, radiation use efficiency and total
biomass production over the annual growth period as demonstrated by Monteith
(1977). Radiation use efficiency is defined as biomass produced per unit of intercepted photosynthetically active radiation (g MJ^{-1}). MISCANMOD was
applied across Europe to predict biomass production of *Miscanthus* (Stampfl, et
al, 2007).

 When the MISCANMOD model was used to explore the likely productivity
of *Miscanthus* in Illinois in the Midwestern USA, projections of peak annual
biomass prior to senescence ranged from 27 to 44 t ha^{-1} (Heaton et al., 2004b).

A meta-analysis of the effects of management factors on *M. x giganteus* growth and biomass production in Europe (Miguez et al., 2008) has shown that one of the simplest models for predicting potential biomass production based on the thermal units accumulated during the growing season provides a remarkably good fit to the observed data. The data indicate that once the normal agronomic practices such as weed control and water availability are in place, temperature accounts for most of the variation in growth patterns.

The critical factors in the development of these mechanistic models to determine yields of *Miscanthus* for different climatic conditions are variation in radiation use efficiency with soil water availability, temperature and drought as well as frost tolerance. Models are being continuously improved to increase their applicability in a wider range of environmental and management scenarios. For example, Hastings et al. (2009a) developed a new model called MISCANFOR that is based on the same processes as MISCANMOD but with improved descriptions of light interception by the canopy and the impact of temperature and water stress on the radiation use efficiency in photosynthesis. Figure 3.3 shows the major data inputs, calculation stages and outputs of this model. They also added genotype specific descriptions for the plant growth phase, photoperiod sensitivity, frost-kill predictions, nutrient repartitioning to the rhizome and moisture content at harvest. When predictions made with MISCANFOR were compared with MISCANMOD for 36 experimental data sets for a wide variety of soil and climate conditions in Europe, MISCANFOR matched field experiments with an r^2 of 0.84 compared with 0.64 for MISCANMOD. Furthermore, the comparisons between the two models using the mean weather conditions for the period 1960 to 1990 in Europe showed that the previous estimates of peak yield made with MISCANMOD had been underestimated in the temperate regions by 20% and grossly overestimated yields in the arid regions. In order to assess the impact of future climate on *Miscanthus* production MISCANFOR has also been run forward for 2020, 2050 and 2080 using IPCC climate projections (Hastings et al., 2008) (Figure 3.4). In order to achieve maximum energy yield with minimum GHG emissions only arable land that is surplus for food production was used for biofuel production.

Robertson et al. (2015) identified five ecosystem models, incorporating carbon cycle dynamics and in particular the soil carbon dynamics, and parameterised for *Miscanthus*. These are: WIMOVAC (Miguez et al., 2009), BioCro (Miguez et al., 2012), Agro-IBIS (VanLoocke et al., 2010), DayCent (Davis et al., 2010) and DNDC (Gopalakrishnan et al., 2012). Of these models, WIMOVAC, BioCro and Agro-IBIS were originally created to simulate biomass production, but subsequently had soil biochemistry and soil carbon transformations incorporated in their simulations. Conversely, DayCent and DNDC were originally designed to simulate belowground nutrient cycling but subsequently more complex plant growth routines were incorporated. The general characteristics of each of the models reviewed by Robertson et al. (2015) are summarised in Table 3.1. The justification for the requirement of the models to simulate carbon dynamics was because assuring the commercial viability of a *Miscanthus*

Table 3.1 Miscanthus mechanistic models – general characteristics

Model name	Spatial scale	Sub-models	Model-type	First reference
MISCANMOD	Field	P	Crop Growth	Clifton-Brown et al. (2000)
MiscanFor	Field	P,W	Crop Growth	Hastings et al. (2000)
WIMOVAC	Ecosystem	P,W,N,C	C/N cycle	Miguez et al.. (2009)
BioCro	Ecosystem	P,W,N,C	C/N cycle	Miguez et al.. (2012)
Agro-IBIS	Region	P,W,N,C	C/N cycle	VanLoocke et al.. (2010)
DAYCENT	Field	P,W,N,C	C/N cycle	Davis et al.. (2010)
DNDC	Field	P,W,N,C	C/N cycle	Gopalakrishnan et al.. (2012)
ECOSSE	Field/ Region	P,W,N,C	C/N cycle	Smith et al.. (2010)

P, plant growth; W, water; N, Nitrogen; C, soil carbon dynamics (soil organic matter included) (adapted from Nair et al. (2012) and Robertson et al. (2015)).

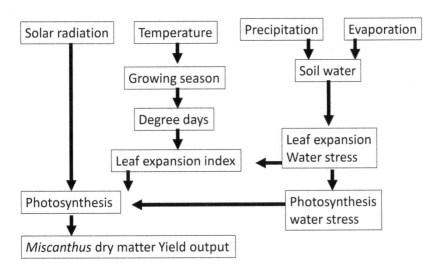

Figure 3.3 Block diagram of the MiscanFor plant growth module showing the major data inputs, calculation stages and outputs
(From Hastings et al., 2008).

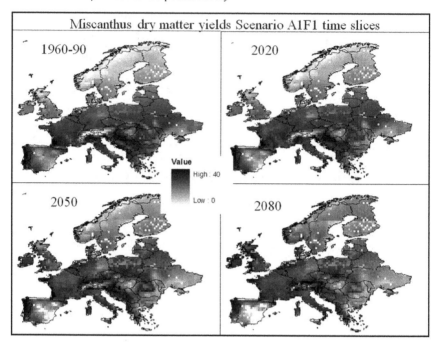

Figure 3.4 Miscanthus predicted yields for the years 2020, 2050 and 2080 compared to a baseline case, which is the average of the period 1960–1990, mapped across Europe using the MiscanFor model. The shade scale is from black (40 Mg ha^{-1} y^{-1}) to white (0 Mg ha^{-1} y^{-1}) of *Miscanthus* dry matter. The climates for the years 2020, 2050 and 2080 were the A1F1, IPCC scenarios
(From Hastings et al., 2008).

plantation and assessing its impacts on greenhouse gas emissions are essential before a landowner can decide whether to establish *Miscanthus* at the expense of other income generating land uses. The Robertson et al. (2015) review of the five process-based crop models parameterised for *Miscanthus* found that they differed both in their design, computational power and spatial scale but none was vastly superior and the main differences were their ability to deal with the specific research questions they were designed for.

Increasingly, the outputs from the ecosystem models are being incorporated into coarser, global-scale integrated assessment models (IAMs) that perform cross-sectoral cost optimisation analyses. The IAMs are a suite of tools developed jointly by scientists and economists to answer central questions about climate change, from how the world could avoid 1.5°C of global warming at the lowest cost, through to the implications of countries' current pledges to cut emissions. In relation to bioenergy crops, this integration enables an assessment of the potential for bioenergy and carbon capture and storage (BECCS) to contribute to GHG mitigation alongside competing energy technologies and other measures – a topic that we return to in Chapters 6 and 7.

4 Environmental sustainability of *Miscanthus*

Introduction

Sustainability is a term that is used widely but is difficult to define and is very dependent on the context. There are three pillars of sustainable development: environmental, economic and social. In essence, sustainability is the process of maintaining change in a balanced fashion, and sustainable development is the development that meets the needs of the present without compromising the ability of future generations to meet their own needs (Helm, 2015). Clearly, ensuring a reliable and sustainable supply of biomass to meet policy targets raises considerable challenges. In an attempt to promote the wider production and use of modern bioenergy the Global Bioenergy Sustainability Partnership (GBEP, 2011; Scarlat and Dallemand, 2011) proposed 24 indicators of sustainability, under the three pillars of sustainable development, intended to inform policy makers and facilitate the sustainable development of bioenergy (see Table 4.1, Gabrielle et al., 2014). Here we can view it in terms of sustainability of the *Miscanthus* crop, the sustainability of the agricultural system incorporating *Miscanthus*, or the sustainability of the whole life cycle of the *Miscanthus* bioenergy system. The social implications of bioenergy production are important to assure the sustainability of biomass based energy production. One of the challenges associated with conducting a social sustainability assessment of bioenergy chains is that the geographical dispersion and heterogeneity of the population potentially affected depends on the legal framework, institutional arrangements, social norms and socio-economic characteristics of the affected population. The complexity of the assessment of the social implications is high and methodologies are still in the development stage.

One of the most important challenges to growing energy crops is to maximise productivity, but doing so in an environmentally sustainable manner. If *Miscanthus* is to become a major feedstock its economic feasibility depends on achieving the highest possible yields, but this cannot be done if it is at the expense of the environment. The aim of long-term management of any biomass crop is to balance the need to maximise net biomass production, while achieving sustainability goals and environmental benefits. Consequently, all bioenergy supply chains must, on the one hand, be as productive as possible and energetically favourable. And here *Miscanthus*, with energy output/input ratios around

Table 4.1 Sustainability indicators of bioenergy use developed by the Global Bioenergy Sustainable Partnership (GBEP) under the three pillars of Environmental, Social and Economic. (Based on Gabrielle et al., 2014.)

Environmental	Social	Economic
1. Life-cycle GHG emissions. 2. Soil quality. 3. Harvest levels of biomass resources. 4. Emissions of non-GHG air pollutants. 5. Water use efficiency. 6. Water quality. 7. Biological diversity of the landscape. 8. Land use and land use change related to bioenergy feedstock production.	9. Allocation and tenure of land for new bioenergy production. 10. Price and supply of a national food basket. 11. Change in income. 12. Jobs in the bioenergy sector. 13. Changes in unpaid time spent by women and children collecting biomass. 14. Bioenergy use to expand access to modern energy services. 15. Changes in mortality and burden of diseases attributable to indoor smoke. 16. Incidence of occupational injury, illness and fatalities.	17. Biomass productivity. 18. Net energy balance. 19. Gross value added. 20. Change in the use of fossil fuels and traditional use of biomass. 21. Training and requalification of the workforce. 22. Energy diversity. 23. Infrastructure and logistics for the distribution of bioenergy. 24. Capacity and flexibility of the use of bioenergy.

(Based on Gabrielle et al., 2014).

10 times those of first generation energy crops (McCalmont et al., 2017a) has a distinct advantage. However, at the same time the energy production must be environmentally sustainable. For energy crops, their contribution to reducing greenhouse gas emissions has become of paramount significance in assessing their contribution to environmental sustainability and, as we will see, bioenergy deployment offers significant potential for climate change mitigation, but may also carry considerable environmental risks.

Threats to biodiversity and water and nutrient cycling must be considered alongside greenhouse gas emissions in assessing environmental sustainability. This conflict of interest has been referred to as the 'food, energy and environment trilemma' by Tilman et al. (2009). In essence, to be acceptable to society, biofuel production strategies must be shown to greatly mitigate GHG emissions without jeopardising food and animal-feed production and have minimal negative effects on the environment.

Greenhouse gas emissions

Creutzig et al. (2015), in assessing the impact of bioenergy production on climate change mitigation, have suggested that focusing on the production of lignocellulosic feedstocks, combined with increased end-use efficiency and improved land carbon-stock management appear to be the most promising options for net reductions in greenhouse gas emissions globally. This assessment is based on analyses that have concluded that large-scale deployment of lignocellulosic feedstock, together with BECCS (bioenergy with carbon capture and storage, see p. 82) could help to keep global warming below 2°C above preindustrial levels, in line with the IPCC, Paris Agreement (2015). However, there are concerns that this could also lead to detrimental climate effects as well as negative impact on ecosystems including loss of biodiversity. Uncertainties in technological development and ambiguity in political decisions render forecasts on deployment levels and climate benefits very difficult and, as we will see, the climate impacts of bioenergy systems are ultimately site and case specific because there is a large dependence on local factors, both biophysical and biogeochemical.

Land use effects

The fundamental issue with bioenergy crops is whether they can realise their potential to replace fossil fuels, and therefore mitigate the GHG emissions of fossil fuels, despite the impact they may have as a result of land use changes for bioenergy production. This can only be achieved if, during their full life cycle, they are able to simultaneously minimise on-site emissions of nitrous oxide (N_2O), associated mainly with fertiliser applications, and stimulate carbon sequestration in the soil. This strategy has been the subject of significant debate, particularly because its effectiveness has been demonstrated to be very variable. Studies have reported life-cycle GHG savings ranging from an 86% reduction to a 93% increase in GHG emissions compared with fossil fuels. There is also concern about indirect land-use change as a result of expansion of biofuel production on agricultural land which displaces food production onto land with high carbon stocks or high conservation value, creating a carbon debt which may take decades to repay (Searchinger et al., 2008). However, land management decisions and the type of land converted to bioenergy crops have variable effects on soil carbon stocks and other GHG emissions which have proved difficult to quantify accurately.

The climate benefits of replacing fossil fuels with biofuels has been strongly disputed by a number of analysts who carry out full life cycle assessments (see p. 52) of GHG emissions with specific biofuels and show that there are no net reductions in GHG emissions and they may even be greater than using fossil fuels. They also argue that GHG emissions associated with land use change to biofuels means that, even when biofuel emissions are lower than fossil fuel emissions, it takes many years to recoup the losses accrued during establishment. When assessing the potential climate benefits of biofuels it is essential therefore to consider the consequences of land use change and fertiliser application on carbon stocks and the emissions of nitrous oxide. Don et al. (2012) reviewed the impact of land-use changes to bioenergy production in Europe on GHG fluxes and soil carbon and showed that, although bioenergy crops are expected to make a considerable contribution to climate change mitigation, there are emissions of CO_2, N_2O and CH_4

associated with the change in land use which will reduce, and in some circumstances totally counterbalance, CO_2 savings of the substituted fossil fuel. Transitions from land uses with large carbon reserves, such as permanent and semi-improved grasslands present a particular difficulty. On cultivation of a semi-improved grassland and establishment of a *Miscanthus* crop in Wales, McCalmont et al. (2017) estimated that it would take up to eight years to replace the loss of carbon from the soil. Hughes et al. (2010) used process-based modelling of the carbon cycle to generate estimates of the payback times for *Miscanthus* cultivated on a global scale and found they were between 20–50 years at high latitudes and less than 30 years in the tropics.

While it is true that in the rush to pursue climate change mitigation strategies, the 'carbon neutrality' of bioenergy may not have been rigorously assessed, in recent years careful analyses have shown that perennial bioenergy crops have a far better potential to deliver significant savings in GHG emissions than the conventional 'first generation' energy crops such as corn, palm oil and oilseed rape. In addition, more recent tighter limits on the level of GHG emission reductions introduced in Europe (Council Corrigendum 2016/0382(COD)) and the USA (110th Congress of the United States 2007) has provided a strong stimulus to transition to perennial bioenergy crops.

Nitrous oxide

Nitrous oxide emissions can make significant contributions to the GHG budget of *Miscanthus* cultivation, particularly when nitrogenous fertilisers are applied. However, emissions are much lower when *Miscanthus* crops are grown with the addition of little or no nitrogenous fertilisers, as is normally the case (Drewer et al., 2012). Emissions of N_2O from soils are the result of the microbial processes of nitrification and denitrification (Smith, 2010). These processes are controlled by soil management, such as tillage and abiotic factors such as climate, soil characteristics including water content. Low N_2O emissions during *Miscanthus* cultivation are a consequence of the high nitrogen use efficiency of the crop and therefore the small amount of N fertilisation required during the crop growing cycle. Furthermore, *Miscanthus* does not require annual tillage so that tillage-induced nitrogen mineralisation is minimised.

Although N_2O emissions make a minor contribution to the GHG balance of *Miscanthus* crops once established, there is evidence that N_2O emissions are elevated during establishment due to denitrification associated with high soil nitrate levels following soil tillage, and in particular the increased decomposition of residues from the previous crop. The evidence is that N_2O emissions from perennial crops strongly depend on the previous land use with the greatest risk of large emissions during crop establishment phase. Holder et al. (2018) measured N_2O emissions during establishment of *Miscanthus* on land that had been converted from temperate grassland and found that, in comparison with the uncultivated pasture which was grazed by sheep, N_2O emissions from the *Miscanthus* plots were 550%–819% higher in the first year and 469%–485% higher in the second year (see figure 4.1). This work has clearly demonstrated that N_2O emissions can make a significant contribution to GHG emissions on land use change and their contribution to total GHG emissions needs to be included in any assessment of GHG balances when establishing *Miscanthus* crops.

Soil carbon

Soils have a large capacity to store and build up carbon stocks, or in other words 'sequester' carbon (Jones and Donnelly, 2004). The term 'carbon sequestration' describes the processes by which atmospheric CO_2 is captured and stored in a long-term reservoir in the soil. Plants, through photosynthesis, are the main agents of biologically capturing and storing soil organic carbon (SOC). The carbon cycle of *Miscanthus* can be divided between above and below ground

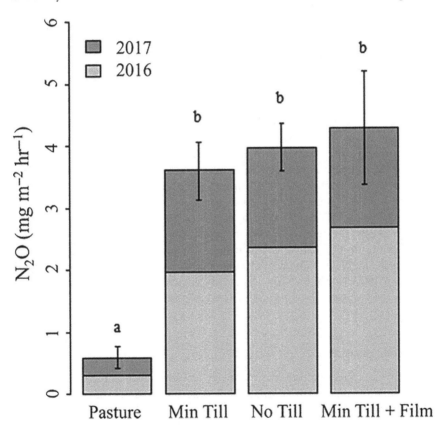

Figure 4.1 Mean cumulative nitrous oxide (N_2O) flux during the first two years after planting a *Miscanthus* crop on formerly grazed grassland on sites that received either no tillage or minimum tillage with or without the addition of a plastic film cover to warm the ground. Error bars show the standard error of the mean and the same letter above the columns indicates non-significant differences (From Holder et al., 2018).

components where the above ground is converted, after harvest, to atmospheric CO_2 either by direct combustion or via a liquid biofuel such as ethanol in the internal combustion engine. Litter and harvest residues undergo a phase of surface decomposition and incorporation. The below ground components enter the soil organic pool dependent on their mortality and longevity. The amount of carbon transferred to the soil is a function of the amount of litter and harvest residues and their decomposition rates, the amount of below ground growth, its longevity and decomposition rates and the depth and distribution of the root system. Agostini et al. (2015) have estimated that the amount of carbon sequestered under *Miscanthus* ranges from about 20–70% of the gross carbon inputs (see Figure 4.2). With increasing age of the crop, the amount of annual SOC increases by more than 40%. The highest rates of SOC sequestration are in arable soils, which are generally low in carbon compared to established grassland. Agostini et al. (2015) conclude that for *Miscanthus* there is evidence for rates of carbon sequestration between 1.4 and 1.88 Mg C ha^{-1} y^{-1}, but they caution that long-term measurements are required to verify sustainable carbon sequestration as the physical and chemical stabilities of the soil organic carbon pool remain uncertain. In another review of 41 literature observations by Zang et al. (2018) they compared conversion of arable land and grasslands to *Miscanthus* and concluded that the average total soil organic matter changes were 6.4 and 0.4 Mg C ha^{-1} y^{-1} respectively (see Figure 4.3). This suggests that establishing *Miscanthus* on low carbon soils, such as degraded cropland, will result in immediate and significant sequestration of soil carbon, whereas converting grasslands *to Miscanthus* will have little benefit in terms of carbon sequestration and may even result in significant short and medium term emissions. The message is that any changes in soil carbon stocks depend critically on the former land-use history and this may dominate the total bioenergy feedstock GHG balance throughout its life cycle. However, a further issue is that soil carbon saturation is likely to occur in the longer term although it is still unclear for most soils where that point is. Dondini et al. (2009), using a modelling approach, projected a steady-state SOC of around 100–110 Mg ha^{-1} for *Miscanthus* growing in Ireland and this value is close to the levels observed in semi-natural *Miscanthus* grasslands in South East Asia.

Ecosystem services

It is widely accepted that in order to be able to assess the net benefits and impacts of perennial biomass crops it is necessary to include more impact categories than greenhouse gas emissions and soil carbon sequestration (Wagner and Lewandowski, 2017). Agricultural use of land produces both commodities such as energy crops and non-marketable goods and services such as soil carbon sequestration, flood control, biodiversity and cultural services. These services are not normally traded but nevertheless they are of great value to society (Helm, 2015). In recent years a substantial research effort has been focussed on the valuation of ecosystem services because they can be used as a basis for decisions that are made for society.

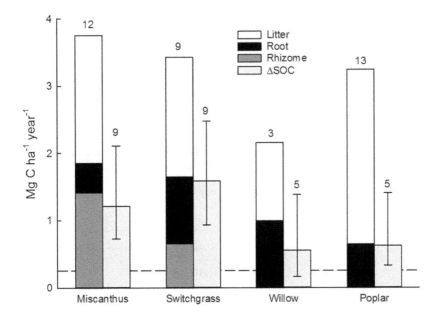

Figure 4.2 Estimates of the annual carbon inputs to the soil from litter, roots, and rhizomes and the change in soil organic carbon (ΔSOC) for four perennial
energy crops. The bars on the ΔSOC column show the maximum and minimum values in the range, and the numbers above all the columns give the
number of values extracted from the literature
(From Agostini et al., 2015).

In order to assess and value ecosystem services, as well as understand the uncertainties underlying the valuation, a strong integration of expertise from both the
natural sciences and economics is required. In economic terms, the aim of this
analysis is to identify land-use strategies that allow the reduction of risks to
ecosystem provision with the smallest loss in expected values of the services.

Land use changes both direct and indirect, and their effects on ecosystem
services are an inevitable consequence of expanding bioenergy production. In
identifying optimal use of bioenergy crops a crucial choice is whether it is
preferable, from an environmentally sustainable standpoint, to cultivate
energy crops with high ecosystem services (land-sharing) or grow crops with
lower ecosystem services but higher yields with the result that less land is
required to meet energy demand (land sparing strategy). To attempt to answer
this question in relation to bioenergy crops, Anderson-Teixeira et al. (2012)
modelled land cover of ecosystem types with their ecosystem services in the
USA. Their conclusion was that the most effective strategy for maximising
bioenergy output as well as ecosystem services was to adopt land sparing and

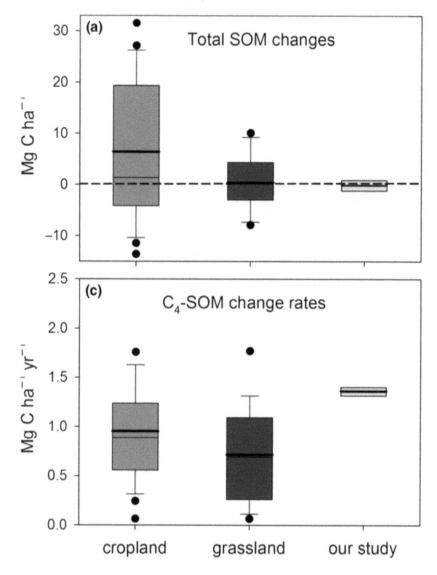

Figure 4.3 Estimates, based on a literature review of 41 observations, of the total organic carbon changes in topsoil (0–20 or 0–30 cm depth) after conversion of former grassland or cropland to *Miscanthus*. The black points are outliers beyond the 10th and 90th percentiles, the boxes are the 25th and 75th percentiles, the central thin line is the median and the central bold line is the median value (Adapted from Zang et al., 2018).

focus simultaneously on maximising yields of bioenergy crops while preserving or restoring natural ecosystems to maintain or increase their biodiversity.

In Europe, where land use is more intensive than in the USA, Milner et al. (2015) assessed the impacts on ecosystem services of land use transitions to second-generation bioenergy crops including *Miscanthus*. This involved the development of a quantitative 'threat matrix' to estimate the potential impacts of transition to these crops, including short rotation coppice, short rotation forestry and *Miscanthus*. The ecosystem effects were found to be largely dependent on previous land use and showed that the complex interplay between land availability, original land use, bioenergy crop type and yield determined the positive and negative impacts of bioenergy on ecosystem services.

In a further European study which compared the ecosystem services impact of first and second generation bioenergy crops, Holland et al. (2015) concluded that there are significant benefits arising from transfer from first generation to second generation bioenergy crops, including disease and pest control and water and soil quality. Figure 4.4 summarises the likely consequences of different transitions for key ecosystem services in the UK and Europe.

Impacts of *Miscanthus* on biodiversity

Biodiversity is the variety of different life forms in natural or managed ecosystems and a high biodiversity is an indication of a healthy ecosystem. Habitat

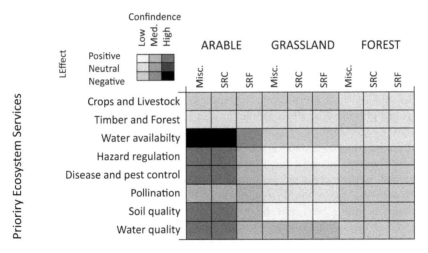

Figure 4.4 Impact matrix, based on a summary of observed effects on priority ecosystem services, of land transitions from arable, grassland and forestry land use to perennial bioenergy crops; *Miscanthus*, short rotation forestry and short rotation coppice. Impacts are scored positive where there is increase in the services, negative with a decrease, and neutral where there is no significant effect reported. Confidence is assigned based on the weight of the evidence (From Holland et al., 2015).

loss will be a major cause of biodiversity decline in the coming 50–100 years and one of the most dramatic examples of biodiversity loss occurs when a diverse community of plants, animals and microbes is replaced by a single species crop grown as a monoculture. When bioenergy plantations displace natural ecosystems they therefore have the potential to inflict serious negative impacts on the environment.

In order to assess the role of bioenergy crops in influencing biodiversity it is vital that the environmental and ecological impacts of their production are measured. In agro-ecosystems, species richness is often correlated with spatial and functional heterogeneity of the environment with a positive influence on biodiversity. In a review of the impacts of bioenergy crops on biodiversity, Dauber and Bolte (2014) conclude that both positive and negative impacts can be observed, depending on a range of factors including the specific location of the production areas, the type of land and land-use shifts involved, the intrinsic biodiversity value and environmental attributes of the energy crop and its associated management practices. They conclude that a clear position about the effects of bioenergy on biodiversity will be very difficult to achieve because of the complexity and potential interactions between these factors.

In the case of *Miscanthus* crops there is evidence that the low agrochemical inputs required for their cultivation, as well as the undisturbed multi-annual growth cycles, result in higher intrinsic biodiversity value of the crop. There is however, concern that areas considered marginal for agricultural production, which often contain habitats of particularly high biodiversity value, will be proposed as localities for *Miscanthus* crop production with negative impacts on biodiversity. Normally, younger crops show greater heterogeneity than more established crops due to gaps between plants and variability in rate of establishment between individuals, but biodiversity is shown to decline as the crop becomes more productive. Dauber et al. (2015) assessed 14 *Miscanthus* fields at the end of their establishment phase (4–5 years after planting) on either improved grassland or tilled arable land in Ireland. Plant species richness and non-crop vegetation cover increased with increasing patchiness due to higher light penetration to ground level. However, the positive effect of *Miscanthus* on biodiversity, measured in terms of numbers of plant and spider species, decreased with increasing productivity. Consequently, they urged caution when assessing the wildlife friendliness of biomass crops as early estimates of biodiversity value may not be maintained at the intensity levels of fully established cropping systems. Furthermore, patchiness has financial implications for farmers, lengthening payback times for initial investment and reducing gross margins significantly (Zimmerman et al., 2014). The progress over time as the crop develops is important when assessing the relative ecological impacts of developing biomass crops in comparison with other land uses. It must also be recognised that there is a trade-off between the goals of climate change mitigation, which depends on high biomass yields and biodiversity conservation within the bioenergy sector.

Bourke et al. (2014) carried out a broad assessment of the response of farmland diversity to the introduction of *Miscanthus* in Ireland. Overall, they confirmed that the cultivation of *Miscanthus* replacing perennial grassland and annual tillage crops had positive effects on species richness of a wide range of taxa. It was concluded that, on the whole, growing perennial crops such as *Miscanthus* is beneficial for biodiversity compared with intensively managed annual tillage crops. This is because perennial crops like *Miscanthus* have long rotation periods, low fertiliser and pesticide requirements, provide better soil protection, offer greater within habitat spatial heterogeneity, have fewer disturbances during the growing season and are harvested in winter.

In an analysis of regional- and national-scale biodiversity data for UK, Haughton et al. (2016) showed that the indicators of weed and insect biodiversity are more abundant in perennial biomass cropping systems than annual arable cropping systems. This study suggests that growing *Miscanthus* supports significantly more biodiversity when compared with arable crops. As a result it was recommended that the strategic planting of perennial biomass crops, such as *Miscanthus*, will increase landscape heterogeneity and lead to higher biodiversity and an enhancement of ecosystem function. However, particular issues arise with any proposals to replace grassland with *Miscanthus* as EU directives have been established in order to protect grassland biodiversity. Under the EU Renewable Energy Directive (RED) an estimated 39–48% (about 9–11 Mha) of natural and 15–54% (about 10–38 Mha) of non-natural grassland in the EU-28 have been classified as highly biodiverse and are not available for conversion to biomass cropping (Hansson et al., 2018).

Invasive potential of *Miscanthus*

Invasive alien species are those non-native species that threaten ecosystems or habitats and are important drivers of human-caused global environmental change through their negative impacts on ecosystem services (Pejchar and Mooney, 2009). *Miscanthus*, like several other potential bioenergy crops, is known to share traits with invasive alien species and has a history of spreading outside its native ranges. Invasion by *Miscanthus* species is probably global but particular concern has been expressed about the escape of *Miscanthus* from cultivation in Europe and the USA (Matlaga et al., 2012). The invasion history of *Miscanthus* species in Europe appears to have similarities in that most populations are small and are unlikely to spread rapidly, although some exceptions of large populations have been identified and might prove difficult to contain. Ornamental varieties have a long history of localised escape in Eastern United States (Quinn et al., 2010). Because the most widely cultivated *M. x giganteus* is sterile it is most likely to invade the margins of biofuel crop fields, where machinery may cause fragmentation of rhizomes (Matlaga and Davis, 2013). However the rate of vegetative spread appears to be rather slow, averaging around 15 cm y^{-1} (Matlaga et al., 2012). Most concern is with the seed producing species, *M. sinensis* and *M. sacchariflorus* and in Europe, *M. sinensis* has been listed on the European and Mediterranean Plant Protection Organisation (EPPO) Alert List of Invasive Alien Plants (https://www.eppo.int/ACTIVITIES/plant_qua rantine/alert_list) since 2011, while *M. sinensis* and *M. sacchariflorus* are on the Watch List in Germany, and restrictions for *M. sinensis* are proposed in Italy. However, at present there are no known planting restrictions on *Miscanthus* planted as energy crops. This contrasts with another PRG, the giant reed (*Arundo donax*), which has been shown to be extremely invasive in the USA and which led to the Environmental Protection Agency introducing regulations that impose requirements aimed at minimising the unintended spread of any Giant Reed intended for bioenergy feedstock production.

Water use and water quality

Agriculture accounts for more than 70% of the world's total water use, and large parts of the world, particularly in the tropics where agriculture is the primary economic activity, already suffer from water scarcity (Coelho et al., 2012). Perennial vegetation can regulate the capture, infiltration, retention and flow of water across the landscape. The vegetation plays a central role in regulating water flow by retaining soil, modifying soil structure and producing plant litter. Water availability in crop plantations depends not only on infiltration and flow, but also on soil moisture retention (Power, 2010). There is no doubt that the production of energy crops for biofuel production will have substantial impacts on water demand, especially if irrigation is used for their production. Production of biomass for energy can, in some cases, expand into areas where conventional food production is not feasible due to water constraints, intensifying water use to the point where stream water or ground water reserves are reduced. It has been estimated that the total water demand for bioenergy production in 2050 could range between 4,000 and 12,000 km^3 per year (Coelho et al., 2012). High-productivity, deep rooted bioenergy plantations of rhizomatous perennial grasses, such as *Miscanthus*, generally have a higher water use than the land cover that they replace even though they are more efficient in their use of water.

Impacts of energy crops on lowering water quality are associated with increases in artificial fertiliser use, nutrient run-off, and erosion. There is evidence that perennial rhizomatous grasses grown in large-scale bioenergy systems can substantially improve the quality of water draining from currently polluted drainage systems. Panagopoulos et al. (2017) assessed the water quality implications of increased biofuel production in the US Corn Belt region. Using a Soil and Water Assessment Tool (SWAT) to model sediment and nutrient losses it was shown that *Miscanthus* along with switchgrass could substantially reduce sediment additions and nutrient levels in the Mississippi River and Gulf of Mexico when grown on the Corn Belt. Thomas et al. (2014) modelled water quality impacts of growing corn, switchgrass and *Miscanthus* on marginal soils in Indiana, USA and simulations showed that switchgrass and *Miscanthus* had no effect on annual run-off but had the positive effects of decreasing water percolation by at least 17% and reducing soil erosion. Furthermore, nitrates leached from fertilised *Miscanthus* plots were approximately 90% less than those leached from fertilised switchgrass and corn systems.

Assessing sustainability using Life Cycle Assessments

Life cycle assessment (LCA) is a computational tool that can be used to evaluate the sustainability of a future biofuel industry. It is a complex tool that lies at the interface between science, engineering and policy. Transparent and accountable LCAs provide a scientific foundation for evaluating the ecological and economic sustainability of biofuels. This holistic view of biofuels is necessary to accurately assess the costs and benefits of alternative fuel systems. Biofuel policies adopted by most countries typically require GHG reduction targets to be met, which are measured through LCA. A full LCA includes cradle-to-grave emissions flows (and/or environmental impacts) starting with biomass cultivation and ending with fuel consumption.

LCA methodology has been increasingly used to assess the potential benefits and/or undesired side effects of biofuels. As we have seen, the conversion of biomass to bioenergy has input and output flows which may affect its overall environmental performances. LCA was developed as a method to compare the environmental profiles of products and services on a 'per-unit' basis (functional unit) and is, in most applications, a static approach. However, the choice of efficiency terms, life-cycle inventories and systems boundaries determines the outcomes of an LCA. Consequently LCA must be carried out with awareness of how each component could influence the outcome. The LCA methodology is regulated by ISO 14040:2006 and ISO 14044:2006 standards which provide the principles, framework, requirements and guidelines for conducting an LCA study, but fundamentally it is a compilation and evaluation of the inputs, outputs and the potential environmental impacts of a product system throughout its life cycle. Despite this standardised approach, the published outputs from LCA studies show a very wide range of uncertainties that currently make it impossible to provide an exact quantification of the environmental impacts of bioenergy crops and energy production. This is largely because of the many variables which can be incorporated into any analysis and also because some of the key parameters (such as indirect effects) are not well known and strongly depend on local and climate conditions (Cherubini and Stromman, 2011).

There are two types of LCA, attributional (ALCA) and consequential (CLCA). ALCA is applied to specific biofuel supply chains, attributing impact to different activities. The EU Renewable Energy directive (European Commission, 2009) bases global warming potential due to greenhouse gas emissions on ALCA calculations. Many sustainability issues are likely to be beyond the scope of the LCA including competitiveness with fossil fuels and food, water security, rural development and health impacts. However, the use of simplified, qualitative scenarios can improve the transparency and insight provided by LCA if uncertainty is acknowledged (Styles et al., 2015). Nevertheless, many LCAs are incomplete because they do not include carbon storage in soils and biomass, and the inclusion of plant, soil and microbial processes that determine greenhouse gas budgets are limited. There is a need therefore, to provide more research data from the plant and soil sciences to improve details of the biofuel production chain (Davis et al., 2008).

Accounting for global net effects of bioenergy production arising particularly from indirect land use change requires a consequential LCA approach. CLCA expands system boundaries to account for marginal effects of system modifications such as displaced food production. CLCA evaluates the potential indirect consequences of biofuels by various 'what if' scenarios. While ALCA is well established, CLCA is still under development and because determining the consequences of indirect land use change is complex, as a result of land transformations and the displacement of crops, there are many uncertainties and possibly misleading outputs.

There have been a number of publications that have carried out LCAs for *Miscanthus* production. Tonini et al. (2012) carried out a consequential LCA of a number of scenarios based on the production of heat and electricity from Danish arable land cultivated with three perennial crops (Ryegrass, Willow and *Miscanthus*). For each crop, four conversion pathways were assessed against a fossil fuel reference. The conversion pathways were anaerobic co-digestion with manure, gasification, combustion in small- to medium-scale combined heat and power (CHP) plants and finally, co-firing in large-scale coal-fired CHP plants. This analysis showed that reductions in GHG emissions compared to the reference only occurred with Willow and *Miscanthus* co-firing.

Wagner et al. (2017) carried out Life Cycle Assessments at six sites in Europe utilising *Miscanthus*. The utilisation pathways used were (i) Small-scale combustion for heat using chips, (ii) Small-scale combustion for heat using pellets, (iii) Large-scale combustion in combined heat and power using bales that were transported and stored, (iv) Large-scale combustion for combined heat and power using pellets, (v) Medium-scale biogas generation for ensiled *Miscanthus*, and (vi) Large-scale production of insulated material. A total of 18 impact categories were assessed. The differences between sites were almost entirely attributable to variation in biomass yields, but the different utilisation pathways varied widely. The analysis highlighted in particular that there are impacts of the different processes on human toxicity, marine and freshwater ecotoxicity and freshwater eutrophication. As a result, in this analysis, the production of insulating material was shown to have the lowest impact on the environment.

An analysis by Murphy et al. (2013) in Ireland incorporated *Miscanthus* cultivation, harvesting, processing and transport to the point of biomass distribution in the LCA assessment. The scenarios examined included replacements of synthetic fertilisers with biosolids, *Miscanthus* processing by pelleting and briquetting and transport distances of 50 and 100 km. This so called 'cradle to gate' LCA study identified hotspots in the *Miscanthus* production chain and indicated that crop maintenance (i.e. fertiliser application) and processing are the stages of the life cycle which contribute most to the environmental impact. It was shown that replacing synthetic fertiliser with biosolids reduced the global warming potential by between 23% and 33% and energy inputs by 12% to 8%, but it was accompanied by large increases in acidification and eutrophication.

Styles et al. (2015) combined the use of farm models and CLCA to compare the net environmental balance of representative biogas, biofuel and biomass scenarios on a large arable farm in the UK and concluded that *Miscanthus* that was pelleted and transported for heating showed advantages over the others because of low inputs, high yields, soil carbon sequestration and ecosystem service benefits. However, the poor financial performance of *Miscanthus* as a crop for farmers, due to low farm gate prices, high establishment costs and a risk premium associated with a 20-year plantation lifetime acts as a major barrier for farm uptake of the crop (Zimmerman et al., 2013). However, Styles et al. (2015) still suggest that incentivisation of *Miscanthus* at the farm level represents better value for money than indiscriminate encouragement of less sustainable bioenergy options via feed-in tariffs and mandatory biofuel blend targets.

Conclusions

As we have seen in this chapter, the production of biomass crops and their utilisation interacts with a host of environmental, ecological, economic and social issues. Environmental impacts include biodiversity and climate through emissions of GHGs and carbon sequestration in soils as well as water availability and quality, and soil and air quality. This has led to a demand for certification of the bioenergy chain encompassing environmental, social and economic aspects; however, the challenge is that the processes and impacts have not yet been fully assessed by the research community (Gabrielle et al., 2014). Table 4.1, however, attempts to summarise the performance criteria that could form the basis of the certification scheme.

Environmental impacts are usually quantified using LCA but their outcomes vary widely. The scale of analysis to provide the inputs for LCAs is also significant so that plot scale trials need to be scaled to regional and global scales in order to achieve applicable conclusions. Assessing the environmental sustainability of bioenergy systems is a wide ranging and complex process which focusses primarily on greenhouse budgets and the provision of ecosystem services. However, the complexity of assessing the large number of energy inputs that go into production of biofuel crops and the extraction of useful energy makes it very difficult to guide policy makers towards identifying the most viable choices and away from costly mishaps. Life Cycle Assessments have been developed to aid policy makers in their decision making, but the different potential combinations of feedstocks, conversion routes, end-use applications and methodological assumptions currently lead to a wide range of, often conflicting, results. The motivation for the promotion of bioenergy and biofuels by governments around the world has been based on the condition that a certain amount of GHG emission savings are achieved. This has meant that standardised GHG accounting procedures, encompassing the inclusion of indirect emissions, have been adopted despite their uncertainties. This formulation of simplistic regulatory standards in the presence of scientific uncertainty may lead to ineffective methodologies. To counteract this, we therefore need continuing research to assess whether the current analyses of renewable energy technologies have adequately quantified energy requirements, outputs and environmental consequences of bioenergy production and utilisation. In particular, the way that indirect effects are estimated urgently needs to be clarified, as they appear to have such a large influence on the final figures (Cherubini and Stromman, 2011).

5   Breeding to improve *Miscanthus*

Introduction

The focus of this chapter is a review of the potential for genetic improvement of *Miscanthus* as a biomass feedstock. While *Miscanthus* hybrids and genotypes, and in particular *M. x giganteus*, have been widely promoted as second generation biomass crops, very little genetic improvement has taken place. Until recently, most emphasis has been placed on improved management to, for example, identify optimal fertiliser requirements and timely harvesting in good conditions to maximise the economic yield of biomass (e.g. Meehan et al., 2013). As we have seen in Chapter 2, the most widely grown *Miscanthus* in productivity trials and in commercial plantations is the sterile hybrid *M. x giganteus*. This hybrid was first collected in Yokohama, Japan by Olsen in 1935 from where it was taken into cultivation in Denmark and subsequently distributed as an ornamental garden plant throughout Europe. The hybrid is probably a result of a cross between *M. sinensis* and *M. sacchariflorus* (Hodkinson et al., 1997). Its selection was based on its ability to both survive better under environmental conditions in Europe but most importantly to out-yield other *Miscanthus* species and hybrids. The prospect of genetic improvement of *Miscanthus* now offers the opportunity to make it better adapted to a wider range of climates and soils than it currently favours and also to produce a crop that can deliver biomass that is optimised to a wide range of end uses.

The three distinct goals associated with development of biofuel feedstocks are: maximising the total amount of biomass produced per hectare per year, sustaining production while minimising inputs, and maximising the amount of energy or bio-based products that can be produced per unit of biomass. Fundamentally, bioenergy crop production should emphasize an optimal balance between input costs and yield, rather than simply maximising yield. The precise values of these parameters will depend on the energy crop and the growing conditions. Research directed towards these goals will likely require the development of systems-level predictive models that will integrate knowledge of molecular and genetic controls with physiological understanding and field crop management. The foundation of this approach will be to use the high through-put tools of genomics, metabolomics and phenomics to rapidly develop the understanding needed to create novel, second generation bioenergy crops. However, it will be important for the molecular biologists to work closely with agronomists to develop genotypes suited to the wide range of field conditions potentially available throughout the world for the cultivation of *Miscanthus*.

Miscanthus taxonomy

Much progress has been made in recent years on the fundamental characterisation of *Miscanthus* species such as their taxonomy and phylogenetics. Taxonomically, *Miscanthus* is a genus of about ten species that are closely related to sugarcane. All species in the genus have a basic chromosome number of 19. Polyploidy is common in the genus and the species form a poyploid complex with diploids, triploids, tetraploids, pentaploids and hexaploids. The genus is native to eastern Asia, southeastern Asia and the South Pacific with the highest species diversity recorded in eastern Asia, especially China and Japan. Its native distribution includes a wide range of climatic zones running latitudinally from temperate south east Russia at 50°N to tropical Polynesia at 20°S. Longitudinally it extends from Burma, Andaman and Nicobar Islands at 92°E. to Fiji at 179°W. (Hodkinson et al., 2015). Hybrids occur naturally within the genus *Miscanthus* itself. These natural hybrids may be sterile because they are triploid. This is the case with *M. x giganteus* which is a hybrid between a diploid (*M. sinensis*) and a tetraploid (*M. sacchariflorus*). It occurs in Japan where the two parent species sometimes grow in the same area (Clifton-Brown et al., 2017, Figure 5.1).

Genetic variation and breeding

As we have seen, the most widely planted genotype is *M. x giganteus* which is a naturally occurring interspecies triploid hybrid between tetraploid *M. sacchariflorus* and diploid *M. sinensis*. However, despite its favourable agronomic characteristics and ability to produce high yields in a range of environments in Europe, the risk of reliance on it as a single clone has been widely recognised. Several recent studies have outlined population genetic variation and examined adaptive variation of a range of genotypes of *Miscanthus*. The ongoing challenge is to combine the genotypic and phenotypic knowledge for crop development and to better incorporate natural genetic diversity into breeding programmes.

Selective breeding of *Miscanthus* for use as an energy crop is still in its infancy compared to food crop species where intensive breeding programmes have been in place for many decades. *Miscanthus* breeding in Europe started in the early 1990s at Tinplant (Magdeburg, Germany) and in Sweden at Svalof Weibull (Clifton-Brown et al., 2008). In 2004 a breeding programme was initiated in the Institute of Biological, Environmental and Rural Sciences in Aberystwyth University, Wales and more recently programmes are underway in the USA and France (Clifton-Brown et al., 2008). All programmes have placed a high priority on collection and characterisation of diverse germplasm. When accessions from the germplasm collection have been planted in nursery plots, huge phenotype diversity is seen in a

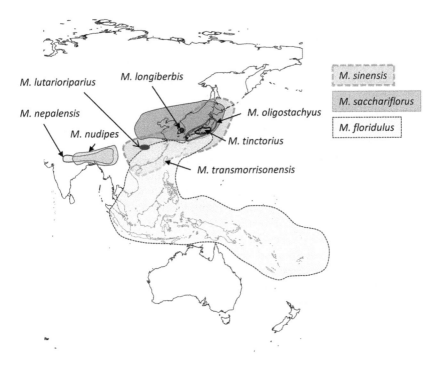

Figure 5.1 Geographical distribution of the major *Miscanthus* species. The distribution of *M. x giganteus* is not fully known, but can be potentially found in regions where *M. sinensis* and *M. sacchariflorus* overlap (sympatric zone) (Adapted from Clifton-Brown et al., 2008).

whole suite of traits which are important for both yield and quality, particularly in *Miscanthus sinensis*, which is the most ubiquitous species in Asia. *Miscanthus sachariflorus* has a more limited geographical distribution than *M. sinensis*, but still shows significant considerable phenotypic variation in characteristics such as stem height and thickness as well as flowering times which are of importance in determining yield.

A starting point for breeding is an assessment of the genetic variability of wild progenitors. Growing plants in what are referred to as 'common gardens', that is under the same climatic and soil conditions, allows breeders to tease apart genetic from environmental effects on the phenotypic variation of important plant traits. An evaluation of three *Miscanthus* species (*M. sinensis*, *M. sacchariflorus* and *M. lutarioriparus*) collected from across their natural distributional ranges in China was carried out by Yan et al. (2012). Ninety three populations were grown in three locations that represented temperate grassland with cold winters; the semi-arid Loess Plateau, and the relatively warm and wet region of Central China. They were assessed for a range of growth traits including plant height, tiller number, tiller diameter, and flowering time. The trials showed that the wild species exhibited a high level of genetic variation in traits that influence crop establishment and production, particularly under less favourable soil and climatic conditions, where the environment is colder and drier than their native habitats. For example, the northern populations of *M. sacchariflorus* had the highest establishment rates at the most northerly sites owing to its strong cold tolerance.

The primary focus of *Miscanthus* breeding programmes is to develop cultivars which are high yielding under a range of climatic conditions but also to provide a feedstock that can be used for a wide range of conversion processes. Different characteristics of species might be exploited in breeding programmes to provide cultivars for specific purposes. For example, *M. x giganteus* and *M. sacchariflorus* have higher lignin content and would appear to be more suitable for thermochemical conversion processes while *M. sinensis* and some *M. x giganteus* clones have lower lignin content and may be more suitable for biochemical conversion processes (Arnoult and Brancourt-Hulmel, 2015) (see Chapter 6).

Genetic improvement in food cereal yields has benefited largely from an increase in harvest index, the ratio of grain to the rest of the above ground dry matter; however bioenergy crop yield improvement depends on increasing **total** above ground production. Increasing total biomass production as opposed to redistribution of biomass is likely to prove a more difficult task which depends largely on increasing the total amount of carbon gain through photosynthesis. As we have seen earlier, *Miscanthus* has C_4 photosynthesis which is already more efficient at using light energy than the more common C_3 photosynthesis and furthermore it is unlikely to benefit from the increasing CO_2 concentrations in the atmosphere as C_4 photosynthesis is already CO_2 saturated (Jones, 2011). Furthermore, *Miscanthus* breeding faces additional challenges which revolve around its slow maturing over the first two or three years from planting and the fact that until recently virtually all *Miscanthus* is propagated vegetatively using rhizomes, or through micro-propagation as the most frequently planted hybrid (*M. x giganteus*) is triploid and does not produce seed.

The goal of plant breeding is the development of genetic variants of crops with new and improved genotypes. This is achieved primarily by the generation of new combinations of existing genetic variation. Plant breeding begins with the evaluation of parents for traits of interest. These can be height, yield, chemical composition, tolerance to drought, heat, or cold, and resistance to pests and diseases. In general, parents are selected for mating because they each have non-complementary attractive characteristics. In an ideal case some of the progeny from these two parents will display some of the favourable traits without compromising the overall performance (Vermerris, 2008). There is a view that energy crops like *Miscanthus* may be harder to improve than food crops. This is due largely to the fact that perennial crops have longer breeding cycles and they produce multiple harvests which need to be assessed in the breeding programme, but also because a breeding target such as nitrogen use efficiency is already high in a C_4 species, so that improvements may only be marginal.

Clifton-Brown et al. (2017) have outlined four major phases in a *Miscanthus* breeding programme. The *first step* is the collection of diverse germplasm with traits that could confer advantages in novel hybrids. The breeding programme requires a broad range of germplasm collected across a range of latitudes and geographies, where target species occur, to maximise the opportunity to capture a full range of trait diversity. However, although many genetic resource collections of *Miscanthus* exist in Europe and the Americas outside of Asia, there is currently neither a directory of *Miscanthus* collections nor a coordinated programme for the conservation of its genetic resources.

The *second phase* is hybridisation. The *Miscanthus* genus is predominantly outbreeding, due to genetic self-incompatibility mechanisms where very low seed numbers are produced via self-pollination. For small quantities of seed, paired crosses are made by bagging together panicles from selected parents (see Figure 5.2). For larger quantities of seed, crosses are made in isolation chambers or in field plots. In either case, a paired cross often results in a seed in both parents. The quantities of seed produced from a cross depends on many factors, including sexual compatibility, flowering synchronicity, humidity, temperature and plant health. Synthetic varieties are used as the main approach to produce varieties that preserve heterozygosity and minimise inbreeding.

Figure 5.2 Hybridisation of *Miscanthus*. Paired crosses are made by bagging together panicles from selected parents
(Courtesy of J.C. Clifton-Brown).

The *third phase* is ex-situ phenotypic characterisation of new hybrids in a range of climates, which is important in understanding genotype x environment interactions. Field evaluations of germplasm in both spaced plots and multi-location trials are used to characterise new accessions for yield potential and chemical composition. All *Miscanthus* species are perennial, so selection of outstanding crosses can only reliably be made after the second growing season or later.

In the *fourth and final phase*, large-scale demonstration trials are used to develop the agronomic practices which are needed to successfully establish, manage and harvest the crop.

Traits for breeding *Miscanthus*

Plant breeding depends on the identification of the most important morphological and/or physiological characteristics or traits that determine the yield and quality of a crop. These characteristics can be used to identify high-yielding clones and breed new interspecific hybrids. For example, Zub et al. (2011) identified key traits determining *Miscanthus* above-ground yield in a field environment for two harvest dates; autumn and winter. Plant height, stem diameter, lateness of panicle emergence or flowering and growth rates were the main traits to be positively related to yield.

In broad terms, some of the most important traits for *Miscanthus* breeding are as follows:

Light use efficiency

Davey et al. (2017) have suggested that light use efficiency should be an important selection criterion in *Miscanthus* breeding programmes for improved yields. Comparing genotypes of M sinensis, M sachariflorus and M. x giganteus with differing canopy architecture they found that the greater yield in M. x giganteus was explained by higher radiation use efficiency and not differences in intercepted radiation or partitioning of photosynthates (see Figure 3.1, Chapter 3).

Flowering time

The flowering time in *Miscanthus* can have a number of effects on the development patterns of the crop and its productivity. Considerable variation in the flowering time of *Miscanthus* genotypes has been found (Jensen et al., 2011). Early flowering, which occurs more frequently in the genotypes of M. *sinensis* than genotypes of M. *sacchariflorus*, shortens the effective length of the growing season, thus reducing the amount of dry matter that these genotypes produce. Where flowering does not occur before the autumn frosts, the onset of senescence and transfer of nutrients below ground is less efficient. For example, at higher latitude sites in particular, M. *x giganteus* does not normally flower before the onset of autumn frosts. This may influence regrowth in the following spring but also increases the ash content in the harvested material which affects the combustion characteristics.

Senescence

Senescence is the process of ageing in plants which, as it develops, leads to the cessation of active growth and ultimately to the initiation of cell death. The process of senescence allows for the more efficient use of nutrient resources by removing nutrients from leaves that are no longer active and moves them and photosynthate into storage organs where they are available for growth in subsequent seasons. There is a genetic basis for variation in the timing and rate of senescence in *Miscanthus* which may be a useful trait for selection of improved germplasm (Robson et al. 2015). For example, 'stay-green' varieties, showing delayed senescence, have the potential for higher yields by extending the growing season but also improving stress tolerance as varieties stay green during a period of water stress and recover soon after the stress is relieved (Clifton-Brown et al., 2002).

In temperate climates such as north-west Europe, there is a strong correlation between growing season length and yield for a number of *Miscanthus* genotypes. Under these conditions the targets for increasing yield has been to select for early leaf emergence in the spring and late flowering/senescence. However these correlations are not so clear in continental climates (Nunn et al., 2017). Here short sharp growing seasons are important and selection focusses on high net photosynthesis and water use efficiency rather than triggers and brakes determining the beginning and end of the effective growing season.

Chemical composition

The optimum composition of the harvested biomass depends on the intended use of the biomass and particular characteristics can present serious limitations on the use of the biomass. For example, if the biomass is combusted, the moisture and mineral content at harvest can strongly influence the efficiency of the process and in some cases can corrode the combustion chamber (Lewandowski and Kicherer, 1997) (see Chapter 6). For use in a range of biorefining applications, cell wall recalcitrance is a major limitation. Recalcitrance refers to the resistance of the cell walls to deconstruction or digestion, which is dictated by relative abundances and interactions between cell wall components (da Costa et al., 2017).

Resistance to pests and diseases

Currently, *Miscanthus* cultivars are proving remarkably resistant to damage by pests and diseases, but as the crop becomes more widely established there will be increasing concerns about the spread of pests and diseases once they become established in certain areas. In Asia, *Miscanthus* plants are often damaged by stem-boring insects (probably lepidopteran stem borers) which use the internodes to breed their young. If the populations of these insects become very abundant then significant yield losses can occur (Clifton-Brown et al., 2008). There are also fears that as the areas of *Miscanthus* increase this may aid the transfer of one of the main aphid-transmitted viruses infecting wheat crops, Barley Yellow Dwarf Virus (BYDV), which has also been identified in *Miscanthus* (Christian et al., 1994). There will clearly be a need to breed resistance to these pests and diseases in the future to avoid the use of expensive and potentially environmentally damaging pesticides and fungicides.

Adaptation to marginal land and stressed environments

Marginal land

As we have seen, in order to avoid competition for land used for food crops it is likely that bioenergy production from *Miscanthus* will utilise marginal land. Marginal lands are those poorly suited to annual food crops because of low crop productivity due to inherent edaphic or climatic limitations or because they are located in areas that are vulnerable to erosion or other environmental risks when cultivated.

Marginal land by definition severely limits the productivity of crops which grow on it as they are subjected to a range of abiotic stresses, including shortage/excess of soil water, low nutrient availability, salinity and high and low temperatures, which reduce their yields below their potential. Plants have evolved functional traits that reflect their ecological strategies in stressful environments so that natural vegetation on marginal land consists predominantly of stress tolerators (Jones & Jones, 1989). Stress tolerant plants normally have inherently low growth rates, so that biomass crops selected for stress tolerance and grown on marginal land might have yields that are so low that they are not economically viable. Stresses on marginal land are often multiple, combining extremes of temperature with water stress and possibly salinity stress and soil toxins. Tolerant plants therefore need to possess combinations of traits that enable them to survive and thrive on marginal land. The objective in exploiting marginal land for bioenergy crops is to identify traits that plants use to avoid or tolerate stress and by using appropriate breeding technology to incorporate these traits in bioenergy crops grown on the marginal land. In broad terms, traits of particular importance on marginal land are tolerance of drought, frosts, low non-freezing temperatures, as well as saline and contaminated soils. One important trait for stress tolerance is the persistent root system of PRGs like *Miscanthus,* which makes them better adapted to the low nutrient status and drought conditions in soils of marginal land. There is however a need to identify other major traits that could be selected for and incorporated into *Miscanthus* in order to utilise the vast areas of marginal land that are available worldwide.

Temperature stress

Low and high temperature stress leads to a number of physiological responses in plants, mainly associated with direct effects on metabolic activity and the process of cell expansion which ultimately control the rate of growth. Extremes of temperature both low and high can be lethal but most attention to date has been given to rapid screening for frost tolerance by, for example, measuring electrolyte leakage following freezing (Perez-Harguindeguy et al., 2013). Since C_4 grasses like *Miscanthus* are of tropical origin, considerable research effort has focused on the influence of low temperatures, including frost and chilling, on growth. Frost tolerance can be divided into two categories: frost tolerance associated with the over-wintering of rhizomes in winter, and frost tolerance of newly emerging shoots in spring. Clifton-Brown and Lewandowski (2000) demonstrated cold tolerance variation between species and genotypes of *Miscanthus* and more recently Yan et al. (2012) found considerable variation in a range of traits including cold tolerance and overwinter survival/establishment among 93 wild collected genotypes of *M. sacchariflorus* and *M. sinensis* gown in common garden experiments at three sites with contrasting climate in China. This confirms that the selection of genotypes with enhanced cold-temperature tolerance represents an effective strategy for maximising production in environments subjected to cold stress.

Others have found that specific leaf area and shoot growth rate may be useful in genotype screening for high productivity and cold tolerance (Jiao et al., 2016). They also found that one *M. sacchariflorus* genotype has higher photosynthetic capacity than *M. x giganteus* at lower temperatures and may be better adapted to cooler climates and can also be incorporated into a breeding programme with *M. x giganteus* to incorporate improved traits for higher growth rates in cool temperate climates. Genotypic variation in the temperature required to initiate growth from the overwintering rhizomes and the effects of frost on the newly emerged leaves in spring were reported by Farrell et al. (2006). Also considerable genotypic variation in the thermal responses of leaf expansion rate of plants already growing was reported by Clifton-Brown and Jones (1997).

It has long been known that low non-freezing (chilling) temperatures, particularly at high light levels, can photo-inhibit photosynthesis in plants with C_4 photosynthesis. However, one genus that appears to be uniquely cold tolerant is *Miscanthus* (Naidu et al., 2003; Naidu and Long, 2004). Chlorophyll fluorescence was used to screen crosses between *M. sinensis* and *M. sacchariflorus* for cold tolerance. Higher tolerance was detected in seedlings from *M. sinensis* mothers, indicating participation of the maternal cytoplasm in the heritability of this trait. There was also a high correlation between maximum leaf photosynthetic rates in chamber-grown *Miscanthus* genotypes and field measurements of shoot growth indicating that this trait is directly related to growth rate in this genus. Furthermore a positive correlation between specific leaf area and leaf photosynthesis was demonstrated, indicating that this trait may be deployed as a screening tool in *Miscanthus* breeding programmes.

More recent screening of a large number (864) of accessions from different populations of *Miscanthus* species and genotypes collected in Japan has identified one particular accession of *M. x giganteus* with a greater capacity for photosynthesis under chilling than the widely grown 'Hornum' clone (Glowaka et al., 2015). This is likely to provide important material for breeding new chilling tolerant *M. x giganteus* genotypes and as suggested by Glowaka et al. (2016) it may even be possible to transfer this cold tolerance to the currently chilling sensitive sugar cane and allow this crop to be grown more widely in temperate climates.

Drought stress

Drought stress tolerance is an important breeding goal in *Miscanthus*. Drought stress tolerance is the ability of plant tissue to maintain metabolism at low plant water potentials brought about by reduced water uptake from drying soil. Selection for drought tolerance has been achieved by selecting for plants that have higher water use efficiencies measured as the ratio of the amount of biomass produced to amount of water lost through evapotranspiration. There is evidence that drought tolerance varies extensively among *Miscanthus* genotypes so that there appears ample scope for breeding drought-tolerant varieties that can still produce substantial yields under water deficits (Malinowska et al., 2017).

Drought stress not only reduces yield it can also alter quality of the biomass. Furthermore, higher degradability of drought treated plants may substantially increase cellulose conversion efficiency in the production of ethanol. However it is still unclear whether in terms of the total ethanol yield per hectare, the reduction in cellulose and biomass yield associated with drought are compensated for by an increase in conversion efficiency. There is evidence that biomass quality characteristics and drought tolerance are largely under independent genetic control. As a result drought tolerance and biomass quality are not mutually exclusive breeding goals and biomass quality can be selected for independently and simultaneously without adversely affecting drought tolerance and vice versa (Van der Weijde et al., 2017).

Salinity stress

Salt stress is an increasing problem for agriculture and is likely to become worse as summer rainfall declines under climate change and more irrigation is used (Chaitanya et al., 2014). Salt tolerance by plants is achieved by either excluding salt from the roots, secreting excess salt uptake or tolerating salt in the tissues, usually by increasing succulence (Perez-Harguindeguy et al., 2013). Each of these forms of tolerance is associated with characteristic physiological traits and in the *Miscanthus* genus large variations in salt tolerance between varieties and within species has been demonstrated (Chen et al., 2017).

Recent progress in breeding *Miscanthus*

Natural genetic diversity is high in the *Miscanthus* polyploidy complex and much progress has already been made in the characterization, evaluation and utilization of these resources so that artificial selection is not restricted by a lack of variation. The natural genetic diversity in *Miscanthus* has been characterised to define gene pools and used to help direct novel crossing work, manipulate ploidy, undertake QTL (quantitative trait loci) and association mapping studies, and develop genomic selection programmes. It has even been suggested that *Miscanthus* can serve as a model for the use of genetic resources for new crop development (Hodkinson et al., 2015). Advances in genetics underlying agronomic traits and the manipulation of these characteristics in breeding programmes will depend on the efficient utilisation of existing collections and also on future collections aimed at targeting a maximum natural genetic diversity. There is therefore a need to develop detailed phenotyping descriptor lists, a network of genetic resource collections and better seed/field bank coordination at the international level (Hodkinson et al., 2015).

As indicated earlier, in breeding bioenergy crops the first priority will be the energy yield per hectare and the energy output:input ratio closely followed by a need to tailor genotypes for the local conditions of climate and soil. However, as the potential for more and varied end-uses becomes clear, then it will be increasingly important to tailor *Miscanthus* genotypes for these end-uses. The important challenge is to identify methods to screen germplasm rapidly for the required traits and to create hybrids from parents displaying these traits.

Germplasm collections specifically to support breeding of *Miscanthus* for biomass started in the late 1980s and early 1990s in Denmark, Germany and the UK (Clifton-Brown et al., 2017b). These collections have continued with successive expeditions from European and US teams assembling diverse collections from a wide geographic range in eastern Asia, including from China, Japan, South Korea, Russia, and Taiwan (Hodkinson et al., 2015). Three key *Miscanthus* species for biomass production are *M. sinensis, M. floridulus* and *M. sacchariflorus*. *M. sinensis* is widely distributed throughout eastern Asia, with an adaptive range from the sub-tropics to southern Russia. This species has small rhizomes and produces many tightly packed shoots forming a 'tuft' morphology. *M. floridulus* has a more southerly adaptive range with a rather similar morphology to *M. sinensis*, but grows taller with thicker stems and is evergreen and less cold tolerant than the other *Miscanthus* species. *M. sacchariflorus* is the most northern-adapted species ranging to 50°N in eastern Russia. Populations of diploid and tetraploid *M. sacchariflorus* are found in China and Korea, and eastern Russia, but only tetraploids have been found in Japan. Germplasm has been assembled from multiple collections, though some early attempts to do this are poorly documented. This historical germplasm was used to initiate breeding programmes largely based on phenotypic and genotypic characterisation. As many of the accessions from these collections are 'origin unknown', crucial environmental data from collection sites is not available. More recently, UK-led expeditions started in 2006 and continued until 2011 with European and Asian partners and have built up a comprehensive collection of 1,500 accessions from 500 sites across Eastern Asia, including China, Japan, South Korea and Taiwan (Clifton-Brown et al., 2018). These collections were informed by using spatial data to identify more detail of the climatic conditions. Accessions from these recent collections were planted, following quarantine, in multi-location nursery trials at several locations in Europe to examine trait expression in different environments. Based on the resulting phenotypic and molecular marker data, several studies characterised patterns of population genetic structure, identified preliminary marker-trait associations and assessed the potential of genomic prediction (Davey et al., 2017).

Separately, US-led expeditions also collected about 1,500 accessions between 2010 and 2014 and produced a comprehensive genetic analysis of the population structure for *M. sinensis* (Clark et al., 2015). Multi-location replicated field trials have also been conducted on these materials in North America and in Asia. To date, about 75% of these recent US led collections are in nursery trials outside the USA which, due to lengthy US quarantine procedures, are not yet available for breeding there. However, molecular analyses have allowed the identification of genotypes that best encompass the genetic variation in each species.

While most *M. sinensis* accessions flower in northern Europe, very few *M. sacchariflorus* accessions flower even in heated glasshouses. For this reason, the European programmes in the UK and Netherlands have performed mainly *M. sinensis* (intraspecies) hybridisations. Selected progeny become the parents of later generations (recurrent selection). Seed sets of up to 400 seed per panicle occur in *M. sinensis*. In Europe and the USA, significant efforts to induce synchronous flowering in *M. sacchariflorus* and *M. sinensis* have been made because interspecies hybrids have proven higher yield performance and wide adaptability (Kalinina et al., 2017). In interspecies pairwise crosses in glasshouses breathable bags and/or large crossing tubes or chambers in which two or more whole plants fit are used for pollination control. Seed sets are lower in bags than in the open air because bags restrict pollen movement whilst increasing temperatures and reducing humidity. About 30% of attempted crosses produced 10 to 60 seeds per bagged panicle. The seed (0.5–0.9 mg) are threshed from the inflorescences and sown into modular trays to produce plug plants, which are then planted in field nurseries to identify key parental combinations.

Recently, conventional breeding in the EU had produced intra- and interspecific seeded hybrids (Clifton-Brown et al., 2018). When a cohort of outstanding crosses have been identified, it has been important to work on related upscaling matters in parallel. These are:

- Assessment of the yield and critical traits in selected hybrids using a network of field trials.
- Efficient cloning of the seed parents.
- High seed production from field crossing trials conducted in locations where flowering in both seed and pollen parents are likely to happen synchronously.
- Scalable threshing methods for different seed sizes and the levels of seed viability.

The results of these parallel activities need to be combined to identify the up-scaling pathway for each hybrid with the aim of achieving commercial viability. The UK-led programme with partners in Italy and Germany show that seed-based multiplication rates of 1:2,000 are achievable for several interspecific hybrids (Clifton-Brown et al., 2018). The multiplication rate of *M. sinensis* is higher, probably 1:5,000–10,000. In comparison, conventional cloning from rhizome is limited to around 1:20, *i.e.* one hectare of rhizome production could supply around 20 hectares of new plantation.

In addition to phenotyping germplasm, molecular techniques currently used by breeders of crop species are likely to accelerate breeding in *Miscanthus*. There have been several efforts to assess variation at the genetic level to inform breeders of certain wide crosses which it would be interesting to attempt. In addition, genetic mapping has been used to study flowering time and certain quality traits. Also molecular markers such as simple sequence repeats, single nucleotide polymorphisms and amplified fragment length polymorphisms (AFLPs) have been used in associating genetic differences with the phenotypes with a view to gene discovery.

More recently, breeding programmes based more on molecular techniques have focussed on developing molecular markers for breeding in the EU, the USA, South Korea and Japan. There are several publications on QTL mapping populations for key traits such as flowering and compositional traits (e.g. Atienza et al., 2003). In the USA and UK, independent and interconnected bi-parental 'mapping' families have been studied (Gifford et al., 2015; Dong et al., 2018) alongside panels of diverse germplasm accessions for genome-wide association mapping studies (GWAS). Further developments have involved calibrating genomic selection with very large panels of parents and cross progeny (Davey et al., 2017) and the completion of the first *Miscanthus* reference genome to improve the efficiency of marker-assisted selection strategies, and especially GWAS.

Robust and effective *in vitro* regeneration systems have been developed for *Miscanthus sinensis, M. x giganteus* and *M. sacchariflorus* (Ślusarkiewicz-Jarzina et al., 2017). However, there is still significant genotype-specificity and these methods need 'in house' optimisation and development to be used routinely. They provide potential routes for rapid clonal propagation and also as a basis for genetic transformation. For *Miscanthus sinensis,* reports exist of stable transformation using both biolistics (Wang et al., 2011) and *Agrobacterium tumefaciens* DNA-delivery methods (Hwang et al., 2014). There are no reports of genome editing in any *Miscanthus* species but new breeding innovations, including genome editing, are urgently needed in this slow-to-breed, non-food, bioenergy crop. It is also likely that developing energy crops through biotechnology would be less controversial than the similar approach in food crops that has met legislative opposition in Europe.

Performance-testing genotypes

Long term field trials to assess the performance of *Miscanthus* genotypes are relatively few in number. Gauder et al. (2012) reported on a 14 year field trial in southwest Germany where 15 genotypes of *Miscanthus* including *M. x giganteus, M. sacchaiflorus* and *M. sinensis*. The best performing genotype was *M. x giganteus* and the highest yield potential was associated with either the absence of or late flowering. In a concerted effort to accelerate the domestication of *Miscanthus* for the UK, Robson et al. (2013) established a productivity trial with 244 genotypes from *M. sinenesis* and *M. sacchaiflorus* hybrids including diploid, triploid and tetraploid genotypes in the largest diversity trial to date. Plants were analysed for morphological traits and biomass yields over three growing seasons following an establishment phase of two years. Yield is a complex trait and as a result no single simple trait explained more than 30% of dry biomass yield. However it was shown that the primary traits defining biomass yields are stem height, diameter and number which together predict 59% of yield. It was concluded that the selection of a few robust morphological traits that predict yield is necessary for selecting the parents and progeny within the crossing cycle.

Multi-location field testing of wild and novel *Miscanthus* hybrids selected by breeding programmes in the Netherlands and the UK has been performed. These trials showed that commercial yields and biomass qualities could be produced in a wide range of climates and soil conditions from the temperate maritime climate of West Wales to the continental climate of East Russia and the Ukraine (Kalinina et al., 2017). Extensive environmental measurements of soil and climate, combined with growth monitoring, have been used by Nunn et al. (2017) to understand abiotic stresses and develop genotype specific scenarios. Phenomics experiments on drought tolerance have been conducted on wild and improved germplasm, and in field trials in Poland and Moldova recently produced interspecific hybrids have achieved 30% higher yields under drought compared to a control *M. x giganteus* crop (Malinowska et al., 2016; van der Weijde et al., 2017).

Intraspecific seeded hybrids of *M. sinensis* produced in the Netherlands and interspecific *M. sacchariflorus x M. sinensis* hybrids produced by the UK-led breeding programme entered yield testing in 2018. Substantial variation in biomass quality for saccharification efficiency (glucose release as percentage of dry matter) and ash content and melting point has already been generated in intraspecific *M. sinensis* hybrids across different environments (van der Weijde et al., 2017). New programmes are planned to establish more than 20 hectares of new inter- and intraspecific seeded hybrids across six European countries. This programme is building the know-how and agronomy needed to transition from small research plots to commercial scale field sites and linking biomass production directly to industrial applications. The biomass produced by hybrids in different locations will be supplied to innovative industrial end-users making a wide range of bio-based products, both for chemicals and energy. In the USA, multi-location yield trials have been initiated in 2018 to evaluate new triploid *M. x giganteus* genotypes developed in the USA.

Taking new genotypes to the farm

Currently, a major bottle-neck in the commercialisation of new hybrids, which are the products of increasingly sophisticated breeding programmes, is the slow and expensive process of multiplying and planting rhizomes to establish near-homogenous fields. If new *Miscanthus* hybrids are, as expected, to play a major role in the provision of perennial lignocellulosic biomass across much of the world as part of lower carbon economies then it will be necessary to develop rapidly and easily multiplied seed or rhizome sources. Currently the main method of propagation of *Miscanthus* is as rhizomes or rhizome-derived plants (Xue et al., 2015). Rhizomes are taken from established plants and are grown on in greenhouses during the winter and spring. This is an expensive method for establishing *Miscanthus* and in some circumstances results in large plant losses during the first growing season due, in particular, to frost damage of rhizomes. Furthermore, the multiplication ratio is very low. Other plant parts can be used for propagation including nodal buds and nodal stem cuttings; but these methods are slow and relatively expensive given the low multiplication ratio. Micropropagation has also been used successfully, but again this is a relatively expensive operation. More recently there has been a concerted effort to propagate from seeds using what is anticipated to be a much cheaper process called CEED (Crop, Expansion, Encapsulation, and Delivery system) developed by New Energy Farms (http://new energyfarms.com/products/ceeds). CEEDS^{TM} are primed plant tissue, encapsulated in growing medium which can contain growth promoters and crop protection products. They can be planted using conventional equipment or by using equipment developed to automate planting, even into min-till and no-till seedbeds.

However, this form of establishment will most likely only be effective where growth rates are high enough to ensure that well developed over-wintering rhizomes can establish to ensure that plants can survive sub-zero winter temperatures.

Conclusions

The three goals of improving the supply of *Miscanthus* as a source of biomass involve maximising the the total amount of biomass produced per hectare, sustaining production while minimising inputs and optomising the quality of the biomass for end-users. To date this has been largely achieved by opportunistic selection of wild genetic resources. However, future advances in improving feedstock to meet the above demands will increasingly come from selective breeding of *Miscanthus* using a combination of conventional and newly developed methodologies. In this chapter we have seen that there have been significant advances in recent years. This has involved the collection of a large number of wild progenitors which are now being used in breeding programmes based on a range of rapidly developing molecular techniques. However, the task of improving long-lived perennials is complex and time-consuming. Nevertheless, the increased focus on breeding programmes for this crop should eventually be rewarded as they produce *Miscanthus* varieties that are targeted at an increasingly wide range of environments and end uses.

6 Commercial uses of *Miscanthus* biomass

Introduction

The bioenergy supply chain links the sustainable production of biomass feedstock and the final fuel/energy and chemical products which have a commercial value. These supply chains are optimised by focussing on lower costs, fewer environmental impacts and more social benefits. The challenge to transition towards the large-scale and sustainable production and use of biofuels and biomass products is enormous, largely because the range of options is so large and the resources are limited. There is therefore a need to optimise to achieve the greatest benefits of replacing fossil fuels by biofuels.

In the bioenergy supply chain, biomass crops can be used either for combustion for the generation of power and heat, released by combustion in furnaces, or as a feedstock for the production of advanced biofuels, mainly bioethanol and biogas (Yue et al., 2014). Advanced biofuels are particularly important as they are arguably the only long-term sustainable solution available for the decarbonisation of the transport sector such as long-haul transport and aviation. Furthermore, advanced biofuels and bioenergy intermediates play an essential role not only for energy use but also increasingly for energy storage. Bioethanol is produced by fermentation of sugars and low molecular weight carbohydrates while biogas is mainly methane produced by a process of anaerobic digestion of fresh biomass. Compared to their petroleum counterparts the major components in the advanced biofuel supply chain are different in structure and property although they may be similar in function. The most significant difference is at the feedstock or raw material acquisition stage. Oil emerges from point sources, has a perennial supply and can be transported over long distances with minimum cost. On the other hand, biomass has a low energy density, is distributed over a large area and is subject to seasonality; resulting in high collection and transportation costs. Consequently the location of biomass processing facilities needs to be optimised in order to reduce the transportation costs and to lower the capital and operational costs (Sims et al., 2006; Yue et al., 2014).

A study of socio-economic benefits indicated that the bioenergy sector in Europe may be worth 360 billion Euro in turnover by 2050 and generate up to 180,000 new permanent jobs. However in order to achieve this, further research and innovation is required to improve the mobilisation of feedstock while reducing costs while at the same time making production facilities more flexible to use different stocks as well as improving the efficiency of conversion processes. In the 2018 International Energy Agency report on renewables (IEA, 2018), bioenergy was given a special focus in recognition of the fact that half of all renewable energy consumed in 2017 came from modern (i.e. excluding traditional uses of biomass) bioenergy, which is four times the contribution of solar photovoltaic and wind combined. Most of the modern bioenergy contributing to final energy consumption provides heat in buildings or for industry. The rest is consumed in the transport sector or for electricity generation. Furthermore, the report suggests that bioenergy will continue to lead the growth in energy consumption over their forecast period from 2018 to 2023.

Scaling up biofuels infrastructure

An important issue in transitioning from fossil fuels to biofuels is the long term availability of sustainable biomass which is able to meet the demands of the biofuel industry. Estimates of the global technical bioenergy potential from dedicated biomass plantations, including perennial rhizomatous grasses, in 2050 span a range of almost an order of magnitude from <50 EJ y^{-1} to >500 EJ y^{-1} (1 EJ = 10^{18} J) (Smith et al., 2014). The technical bioenergy potential is the amount of the theoretical bioenergy output obtained by full implementation of demonstrated technologies or practices. The reason for the very wide range of estimates relates to the uncertainties in assessing the areas deemed available for bioenergy production and the projected yields per unit land area. Within Europe a report published by the EU Directorate-General for Research and Innovation suggested a sustainable biomass potential of ~500 million tons by 2020 but with appropriate research and innovation measures an increase of up to 120% of this amount could be achieved by 2050 (European Commission, 2016). In order to meet the wide range of national and international climate objectives (European Commission, 2017a) a number of interrelated limiting factors need to be addressed. Firstly, the feedstock of biomass supplies need to be scaled up significantly. Agricultural producers need to extensively invest in production of new lignocellulosic crops and the processes of harvesting, conditioning and transportation of feedstocks. In addition, developers of advanced conversion techniques need to invest in research and innovation and develop technologies for mass production and integrate these into complex biorefineries. Currently, the most advanced biofuel technologies are commercially immature and feedstock supply is limited. However, since the early 2000s bioenergy production in Europe has grown steeply as a result of the implementation of EU and nationally binding renewable targets set for 2020, although the penetration of advanced biofuels has been minor. Despite this, feedstock production will need to be scaled-up further in the mid and long-term in order to assist in meeting the EU 2030 and 2050 climate objectives.

Beyond the farm gate, there are additional significant barriers to bioenergy development which will clearly influence the demand for *Miscanthus*. For example, in the UK, a review by Adams et al. (2011) found that inhibitors to the UK bioenergy development have led to failures in a number of initiatives. The main reasons for failure to develop planned bioenergy plants were found to be: (i) financial problems during the operational lifespan of the plants; (ii) increased transport around bioenergy plants; (iii) delays in land planning approval; (iv) location of the bioenergy plant – visual impacts; (v) mistrust between local community, developers and agencies; (vi) other environmental impacts; (vii) technical problems associated with the conversion techniques.

It is clear that the projected rapid growth in the demand for lignocellulosic bioenergy will require major changes in the supply chain infrastructure for bioenergy crops. Second generation lignocellulosic bioenergy feedstocks have considerably lower bulk densities than first generation grains, resulting in significant logistical challenges for transport and storage. It has been estimated that, even with densification and processing at the site of production, the transport volume by mid-century could exceed the combined capacity of current agriculture and energy supply chains, including grain, petroleum and coal (Richard, 2010). Efficient supply chains could be achieved by decentralising the conversion processes that facilitate local sourcing, satellite processing and densification for long-distance transport. There are possibly three distinct operational models for biomass feedstock supply; independent local suppliers, large contiguous plantations and regional or global commodity markets. It is suggested that local supplies work well for smaller biomass energy facilities including combined heat and power plants. However, the large plantation approach most likely requires a single company to control a large area of contiguous land which they own, while the commodity biomass market would parallel the trading operations for other agricultural commodities such as grain and livestock. Richard (2010) has shown that the commodity system works most efficiently when processing and transport costs are low, so that distances between buyers and sellers have little impact on prices.

Market opportunities for *Miscanthus*

Given its potential as a high yielding, low input lignocellulosic feedstock, there is growing interest in the use of *Miscanthus* biomass for a wide range of applications and in particular in the production of bioenergy and biofuels (Brosse et al., 2012; Van Der Weijde et al., 2016). The four most important bioenergy conversion routes are direct combustion, anaerobic digestion to produce biomethane, enzymatic saccharification and fermentation to produce bioethanol, and pyrolysis. The most common current use of *Miscanthus* involves harvesting in winter after it has senesced and partly dried and then compressing into pellets for combustion. Conversion technologies available for the production of advanced biofuels from *Miscanthus*, most of which include conversion steps which are still in the pilot and demonstration phase, are reviewed below.

Combustion

Miscanthus can be burnt in open fires, in stoves and boilers or in power stations to produce heat, electricity or both heat and electricity (combined heat and power) (Caslin et al., 2011). There are also several forms in which *Miscanthus* can be combusted, including whole bales, chips, pellets and briquettes. In power stations, *Miscanthus* is usually co-fired with coal or peat.

Biomass co-firing has the potential to reduce fossil GHG emissions from coal-fired power stations without substantially increasing costs or infrastructure investments. Furthermore co-firing can be delivered relatively quickly compared with new-build renewable projects. Viewpoints differ on whether co-firing should have a role in decarbonising electricity supply. While it may deliver benefits in reduction of fossil fuel greenhouse gas emissions it can be viewed as encouraging the continued use of coal for electricity generation. When implemented, even at relatively low biomass to coal ratios, there are significant reductions in fossil energy consumption, and solid waste generation, as well as reduced emissions. However, when the proportion of biomass is increased, the nature and chemical makeup of biomass fuels can lead to significant cost increases and maintenance problems due to boiler slagging and fouling issues, increased boiler corrosion, and decreased efficiency.

The process of utilisation of biomass in combustion systems is strongly influenced by the physical and chemical characteristics of the harvested crop. These characteristics include moisture content, bulk density, physical dimensions and size distribution. Biomass heating value is tightly connected with the elemental composition and is affected by the variation in cell wall composition and ash. The reported heating value of *Miscanthus x giganteus* ranges from 17 to 20 MJ kg^{-1} (Brosse et al., 2012). *Miscanthus* biomass at harvest has a low bulk density and variable moisture content which can make its use for combustion purposes difficult. Pelleting upgrades the fuel and facilitates utilisation by increasing bulk density, stabilising moisture content and reducing dust emissions in the handling of the fuels with the result that it leads to a more stable, uniform product with more efficient combustion control.

Chemical elemental content of *Miscanthus,* such as nitrogen, chlorine and sulphur, and also the content of volatiles, ash, and metals, all have an impact on corrosion problems and/or formation of deposits in the boiler during combustion. The volatile content of biomass varies between 70% and 80% on a dry matter basis, whereas for bituminous coal it is about 18%. The chlorine sulphur and metal contents of biomass fuels are responsible for ash-related problems, such as slagging and fouling, corrosion and agglomeration. Corrosion may occur on metallic tube surfaces in the boilers as a result of the chlorine content in the depositions. The degree of fouling and slagging varies with the fuel characteristics, local gas temperatures, tube temperatures and local heat fluxes on each particle (Carroll and Finnan, 2012).

There are particular concerns about the combustion quality of *Miscanthus*. Mineral content including potassium (K) and chlorine (Cl) plays an important role in affecting biomass combustion quality. This is because K and Cl enrichment can reduce ash melting point and cause corrosion issues. The ideal crop characteristics at harvest are low K and Cl to reduce corrosion of boilers, low moisture to reduce spoilage and transportation costs, and low silica (Si) and ash to reduce slagging and consequential downtime (Jensen et al., 2016). In general, early flowering and senescence have been demonstrated to improve overall combustion quality as K, Cl, moisture and ash are reduced, so the recommendation is that late senescing genotypes should be avoided for thermochemical conversion (Baxter et al., 2014). Meehan et al. (2013) have demonstrated that delaying the harvest date and leaving the crop in the field after cutting lowered the chlorine content. Analysis of *Miscanthus* combustion properties show that stems have a better fuel quality than leaves due to their lower ash contents. Application of N and K during the growth of the crop increases yields but elevates the content of these elements in the plant. As we have seen, fouling can be an issue for combused *Miscanthus* and this can be worsened when fertilisers are added to improve yield.

Enzymatic hydrolysis

To produce biofuels like ethanol, the lignocellulosic biomass of *Miscanthus* needs to be pre-treated and enzymatically hydrolysed to fermentable sugars before fermentation by microorganisms. The production of bioethanol and other fermentation products from *Miscanthus* biomass is a multistep process that normally includes pre-treatments, enzymatic hydrolysis and yeast fermentation. Pre-treatment aids the recovery of cellulosic content from lignocellulosic biomass and renders it digestible to cellulases. The goals of pretreatment are (i) the production of highly digestible solids with enhanced sugar yields during enzyme hydrolysis, (ii) the avoidance of degradation of sugars to furans derivatives and carboxylic acid, which act as fermentation inhibitors, (iii) to recover lignin for conversion into more valuable co-products, and (iv) to be cost effective for future commercial scaling. In enzymatic hydrolysis the structure of lignocellulose is first broken down to cellulose, hemicellulose and lignin. Subsequently the hydrolysis process is carried out by cellulase enzymes which are produced by bacteria or fungi. Cellulase is comprised of three different enzymes which attach to lignocellulosic feedstock and convert cellulose to reducing sugars which are then converted to ethanol by fermentation. The efficiency of ethanol production is about 70% and is considered to be economically competitive, particularly if co-production of lignin and xylan-based products add value (Lee and Kuan, 2015). The efficiency of the conversion routes can be improved by optimising the chemical composition and physical structure of the cell walls. Improved quality can be achieved by breeding for these quality traits. For example, in *Miscanthus sinensis* genotypes, cellulose content ranges from

~26 to 47%, while hemicellulose content ranges from ~23 to 43% and lignin content ranges from ~5 to 15% of dry matter. In addition, cellulose content is affected by climate conditions, and harvest time (Lee and Kuan, 2015).

Currently, the total cost of bioethanol production from lignocellulosic feedstock is still much higher than production from starch-based or sugar-based feedstock. Production of cellulosic biofuels requires a larger investment in more diverse enzymes to convert plant cell walls to sugar than is needed to release sugar from starch. Although enzyme costs have decreased in the last few years, the high cost of enzyme production and the requirement for higher enzyme dosage for hydrolysis of biomass are the main hurdles for the economic viability of lignocellulosic bioethanol (Menon and Rao, 2012). Whereas enzymes account for 4.5% of the cost to make ethanol from corn-starch, they account for 17% to 20% of the cost to make ethanol from lignocellulosic biomass. For cellulosic biofuel to be competitive with fossil fuels it is estimated that the cost of enzymes must amount to only 8% to 10% of the total cost, a two-fold reduction from present costs (Shrestha et al., 2015).

In addition to cost, enzyme diversity is an issue because the plant cell wall, with its many polysaccharides, is far more complex than starch. To make full use of plant cell walls, cocktails of enzymes capable of synchronised digestion of the polymers are needed. In order to identify these enzymes Shrestha et al. (2015) have analysed 30 of the fungi most commonly found on decaying *Miscanthus* leaves and have identified fungi that are significantly superior to the most widely used industrial bioconversion fungus, *Trichoderma reesci*.

While these technical problems of scaling up cellulosic bioethanol remain, there is uncertainty, particularly in the USA where there is a plentiful supply of ethanol made from corn, about whether lignocellulosic bioethanol has a viable future (Service, 2010). However, utilising by-products of the bioethanol process can provide a significant additional financial advantage. A particularly valuable by-product of the treatment process is lignin with applications as a polymer modifier, an adhesive, a resin, and more. In addition to lignin, hemicellulose removed in the pretreatment step can be used as a starting material for the production of xylitol and xyligosaccharides. Figure 6.1 shows the pathway for production of bioethanol and value-added by-products from *Miscanthus* biomass.

In order to assess the environmental impacts of ethanol production from *Miscanthus*, Lask et al. (2018) carried out a life cycle assessment comparing the production pathways at two European sites. The sites were: land which is classified as marginal in Aberystwyth, Wales and a better quality, average-yield, site in Stuttgart, Germany. The conclusion was that the environmental impacts of production in Stuttgart were significantly lower than in Aberystwyth, largely due to differences in yield (9.75 Mg ha^{-1} y^{-1} versus 15.32 Mg ha^{-1} y^{-1}). Nevertheless, in both cases the study showed that production of *Miscanthus* ethanol has potential for the reduction of GHG emissions in the transport sector, although the amount of savings depends strongly on site-specific factors such as biomass yield and soil carbon storage. In fact, for the marginal site, carbon sequestration makes a major contribution to meeting the eligibility conditions for financial support under the EU, Renewable Energy Directives.

Anaerobic digestion – biogas

Anaerobic digestion is the decomposition of organic matter in an anaerobic (oxygen-free) environment to produce biogas that is usually made up of around 60% methane and 40% CO_2. The process of anaerobic digestion involves the production of biogas from fatty acids, amino acids and sugars. Through fermentation or acidogenesis sugars are further broken down into simpler molecules by the action of bacteria in the digestate. Ammonia, hydrogen, carbon dioxide and hydrogen sulphide are the main products of this process. The products are further processed during acetogenesis when acetic acid, carbon dioxide and water are formed. Methanogens then act on these products to form methane. The technology of biogas production is well established and can be implemented in simple set-ups for single households or small communities in developing countries or as part of bigger complex plants feeding a larger grid system (Frydendal-Nielsen et al., 2017).

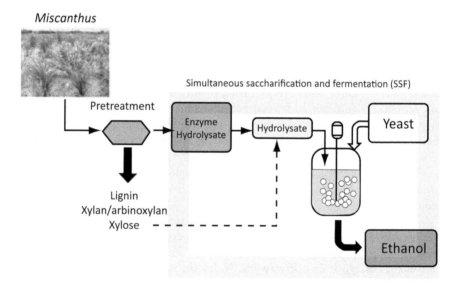

Figure 6.1 Production of bioethanol and value-added by-products from *Miscanthus* biomass. After pre-treatment, the cellulosic fraction of *Miscanthus* biomass is converted by enzyme hydrolysis into simple sugars in the hydrolysate which are then fermented to yield ethanol. Value added products, including xylitol, xylooligosaccharides and lignin can also be co-produced from the matter removed in the pre-treatment step
(Based on Lee and Kuan, 2015).

In Germany it has been suggested that *Miscanthus* has the potential to increase the sustainability of the anaerobic digestion sector by replacing a substantial area of maize cultivation for biogas production (Kiesel and Lewandowski, 2017). Here the *Miscanthus* is cut green, late in the growing season in October, but before significant senescence occurs. The replacement of maize by *Miscanthus* for biogas generation offers great potential for reducing the environmental impact of biogas production without increasing land-use competition as currently maize has a negative environmental impact due to soil erosion, nitrate leaching risks and a negative impact on biodiversity due to the requirement for pesticide applications (Purdy et al., 2017). *Miscanthus* therefore appears to be a very promising crop for biogas generation as it offers the potential to improve the sustainability of the anaerobic digestion sector, particularly in Germany, by replacing a substantial area of biogas maize cultivation. Despite these clear advantages over maize, *Miscanthus* is currently very little used for feedstock in biogas plants probably because of uncertainty about the optimal harvest regime and the resilience of the crop to green cutting. However, recent research by Mangold et al. (2018) has confirmed that *Miscanthus* should be harvested in October to maximise yields and nutrient recycling and that the focus on future breeding of varieties for biogas production should be on higher proportions of leaf to stem.

Pyrolysis

Pyrolysis is the thermal decomposition of biomass in the absence of air. There are a large number of pyrolysis technologies that have been developed with a focus on targeted final products which can be divided into (i) biochar and heat, (ii) biochar, bio-oil and gases, (iii) biochar, carbon black, and syngas (gas mixtures that contain varying amounts of CO and H) and (iv) syngas (Peláez-Samaniego et al., 2008).

Biochar is a carbon-rich solid, formed by what is referred to as slow pyrolysis of biomass at temperatures around 400°C. It can be utilised as an absorbent and/or a soil amendment. It is increasingly recognised as a soil conditioner because its incorporation into the soil has been shown to enhance nutrient balance by supplying and, probably more importantly, retaining nutrients. The unique properties of biochar in the soil include low density (providing aeration in the soil), significant adsorption and cation exchange capacity, and the ability to encourage microbial activity in the soil. Its storage in soils has been suggested as an important means of abating climate change by sequestering the carbon produced by the pyrolysis of biomass (Smith, 2016). Biochar is generally considered a win-win solution in terms of sustainable land use, as in comparison to direct combustion, the conversion of large quantities of organic matter into stable carbon pools by pyrolysis creates a long-term sink for carbon which has been captured from the air in the process of photosynthesis. Persistence in soils is a fundamental quality of biochar in serving its role in carbon sequestration but this persistence should exceed 100 years to match the definition of permanent removal. Biochar made from *Miscanthus* straw has been demonstrated to be very effective compared with other sources such as coffee husks and woody material from woodchip production (Houben et al., 2013), and there is experimental evidence that it is highly persistent in soils, decomposing at an estimated rate of only 0.8% per year (Rasse et al., 2017).

Compared with slow pyrolysis to produce biochar, fast pyrolysis occurs at higher temperatures of around 500°C and a short hot-vapour residence time of less than two seconds. The products of fast pyrolysis of biomass are alkenes, alkanes, aromatic compounds, esters CO_2, CO, water and H_2. On cooling a dark brown homogeneous liquid called bio-oil is produced, and biochar is a by-product.

When the main product is bio-oil, yields up to 75 wt. % on a dry feed basis are obtained. There are multiple ways by which bio-oil can be catalytically up-graded to final liquid fuel products (Bridgewater, 2012). Bok et al. (2013) assessed product yields and bio-oil quality of *Miscanthus sinensis* in fluidised bed reactors and achieved the highest yields of up to 50% in the cylindrical reactor at 450°C. The characteristics of the biocrude oil from *M. sinensis* were similar to the widely known characteristics of that obtained from woody biomass. It was concluded that biocrude oil from *M. sinensis* can be used as a renewable liquid fuel for a range of applications, including gas turbines and diesel engines (Bok et al., 2013).

The process of gasification of biomass to produce bio-syngas is done in two types of gasifiers, either fixed/moving bed or fluidised bed. Fixed bed gasifiers have a simpler construction and require minimum pre-treatment of feedstock. There are four separate zones: drying, pyrolysis, combustion and reduction. In fluidised bed reactors, feedstock, air, oxygen and steam enter from the bottom of the reactor. It has a uniform temperature in the gasifier and the drying, pyrolysis and oxidation reactions occur in the same zone. The hot fluidising gases, the hot sand bed, and the feedstock are mixed intensively inside the reactor to achieve production of bio-syngas.

Non-energy uses

Currently, while the energy utilisation technologies are developing there are other possible markets for *Miscanthus*, such as use for animal bedding and in particular for horse bedding (Rauscher and Lewandowski, 2016), in the manufacture of medium density fibre board (MDF) (Visser and Pignatelli, 2001), for thatching, for paper production (Cappelleto et al., 2000), and as a horticultural material for the construction of plant pots. However, although it has been shown in all cases that *Miscanthus* could be utilised, and may even have some minor advantages, the cost of producing *Miscanthus* is considerably higher than the cost of producing most of the biomaterials currently used. Under the circumstances it is very unlikely that *Miscanthus* will provide a viable feedstock for these processes, at least until the cost of production falls to the point where it is cost competitive.

Biorefineries

A biorefinery is a facility that integrates biomass conversion processes to produce biofuels and value-added chemicals from biomass. It is analogous to a petroleum refinery which produces multiple fuels and products from petroleum. In this process a raw material consisting primarily of renewable polysaccharides and lignin enters the biorefinery through an array of processes; it is fractionated and converted into a mixture of products, mainly of transport fuels, but about 5% goes to high value chemical products (Ragauskas et al., 2006). Bioprocessing facilities start with the breakdown of a biomass feedstock by enzymatic hydrolysis of cellulosic fractions to fermentable sugars. The biological fermentation process converts sugars into biobased intermediates that can be refined into a range of biochemicals, pharmaceuticals and foods. Fatty acids, sterols and other aromatic compounds can also be extracted from *Miscanthus*. The exploitation of these low-volume-high-value chemicals can make an important contribution to achieving an economic return on growing biomass. Figure 6.2 is an illustration of a biorefinery chain from biomass inputs to the production of fuels, chemicals and bio-materials.

Despite the numerous biorefining applications use of lignocellulosic biomass still remains largely untapped as a consequence of cell wall recalcitrance – resistance of the cell wall to destruction. Compared to starch-based feedstocks, *Miscanthus* is highly recalcitrant and therefore more difficult to process; however, there have been assessments of methods to evaluate its digestibility and it is anticipated that both bioethanol and biogas would be important bulk products in the biorefining process (Frydendal-Nielsen et al., 2017). However, a major difficulty in relation to utilisation is that the same lignocellulosic feedstock may present widely distinct biorefining potentials, depending on the plant's developmental stage and relative organ contribution to the total harvested biomass.

Research aimed at improving lignocellulosic biomass for biorefining has motivated considerable advances in our understanding of cell walls. Identification of desirable cell wall characteristics and the development of crops containing such features are crucial steps for lignocellulosic biorefining optimisation. As cell walls vary in composition it is difficult to assess the quality of a given feedstock in relation to others. In an attempt to overcome this problem, da Costa et al. (2017) have performed a multidimensional cell wall analysis to generate a reference profile for leaf and stem biomass from several *Miscanthus* genotypes harvested at different developmentally distinct time points. This has provided relevant information for the tailoring of *Miscanthus* varieties with desirable characteristics for conversion to biofuels and other biomaterials.

Biorefinery Concept

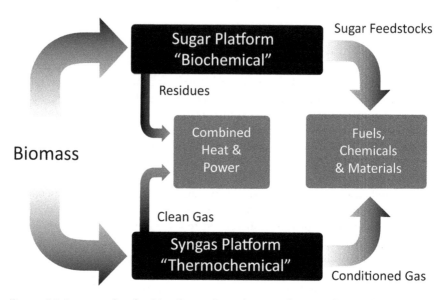

Figure 6.2 An example of a biorefinery chain that uses biomass feedstock to produce fuels, chemicals and materials

Biomass and Negative Emissions Technologies – The future for Bioenergy with Carbon Capture and Storage

The Paris Agreement of December 2015 committed the Contracting Partners to the UN Framework on Climate Change to act towards the objectives of keeping a global temperature rise this century well below 2°C above pre-industrial levels (UNFCCC, 2015). Warming results from increase in GHG concentration in the atmosphere, and in particular CO_2, so reducing CO_2 emissions is recognised as the primary route to achieving these ambitious targets. However, estimates of the impact of decarbonising through radical emissions reductions show that there is a significant gap between the current trajectory in global emissions and the pathway necessary to avoid dangerous climate change (Anderson and Peters, 2016). Consequently, it will be necessary to actively remove CO_2 from the atmosphere ('negative emissions'), in addition to achieving near zero emissions in order to meet the global targets for slowing global warming. The suggested mechanisms for achieving this are referred to as Negative Emissions Technologies (NETs). Currently, global fossil fuel use emits about 8 Pg C y^{-1} to the atmosphere. The oceans and terrestrial biosphere take up approximately 55% of these emissions so that the atmospheric CO_2 concentration grows at about 2 ppm y^{-1}. To slow or reverse this build-up, a number of strategies for atmospheric CO_2 removal (CDR) are being actively considered in order to mitigate climate change. One of the currently favoured CDR approaches involves growing bioenergy, extracting energy from it, and capturing the by-product of CO_2 and storing it underground in geological reservoirs. The whole process is referred to as biomass energy with carbon capture and storage, or BECCS for short (Figure 6.3). BECCS is heavily relied on in scenarios of future emissions in integrated assessment models that are consistent with limiting global mean temperature increase to 1.5°C or 2°C above pre-industrial levels, as was agreed under the Paris Agreement. It is argued that, in principle, BECCS can deliver net CO_2 removal at a global scale sufficient to impact atmospheric CO_2 concentrations and global average temperatures. However, net negative emissions can only be achieved when the CO_2 stored from BECCS exceeds the sum of all systems losses, emissions associated with land use change and emissions from all other human sources, e.g. energy, transport and agriculture. BECCS features prominently in mitigation scenarios for two particular reasons. The first is that it offers the opportunity to delay some of the requirement for emissions reductions in the near future and secondly, the negative emissions from BECCS can compensate for hard to reduce residual emissions in other sectors, particularly transport and agriculture (Smith et al., 2016). To date there has been very little practical experience of implementing the technology that, clearly, still presents a

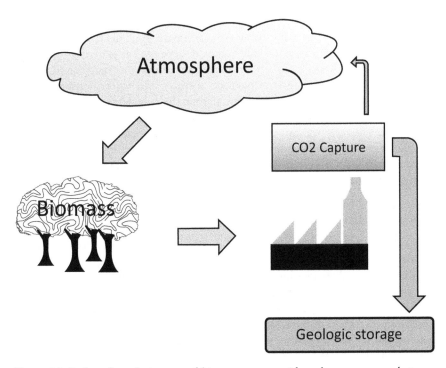

Figure 6.3 Carbon flow during use of biomass energy with carbon capture and storage (BECCS), integrating biomass in heat and power generation sector

number of technical and scientific challenges, including increased N_2O emissions on expansion of bioenergy and the potential leakage of CO_2 which is stored in deep geologic reservoirs (Gough et al., 2018).

BECCS has been referred to as 'benign geo-engineering' as it has the potential to facilitate the transition to a carbon neutral society by offsetting emission sources that are more difficult or more expensive to abate, such as transport. However, although biomass energy is already used in power generation, carbon capture and storage technology is still in its infancy. The challenge is to bring the two together at a sufficient scale to achieve significant negative emissions. A key requirement is to determine the biomass resources required to meet the demand from the biomass-burning power stations (Vaughan et al., 2018). Furthermore, any substantive uptake of biomass for power generation must take into account competing uses for biomass resources from a rapidly developing bioeconomy (Staffas et al., 2013). Smith and Torn (2013) suggest that BECCS faces ecological constraints at the local project level and at larger scales. It is suggested that, although very high sequestration potentials in BECCs have been reported, there has been little systematic analysis of the potential ecological limits to and environmental impact of implementation at the scale relevant to climate change mitigation. Smith and Torn (2013) argue that while it is apparent that BECCS shows obvious promise for climate and ancillary benefits, global scale implementation may be constrained by ecological factors which may compromise the ultimate goal of achieving climate change mitigation. Carbon dioxide removal depends on growing more biomass, which requires land, either currently arable or marginal/degraded land, and this will put pressure on food security and forest conservation in particular (EASAC, 2018). It is essential that the bioenergy resource used is sustainable and has minimal GHG emissions from direct and indirect land use change. However the ranges and magnitudes of associated impacts are hard to quantify due to a combination of insufficient ecological knowledge, unaccounted for GHG emissions for the land use change, as well as the political and economic complexity of the proposed developments.

Conclusions from industry

In this chapter I have reviewed the range of potential commercial opportunities for the use of *Miscanthus*. The Biofuture Platform (BfP) is an example of a government led multi-stakeholder initiative designed to promote international coordination on advanced low carbon fuels and bioeconomy development (Biofuture Platform, 2018). Underlying the BfP commitment is the acknowledgement that an increased penetration of biomass in the energy and materials sectors is essential in enabling the achievement of the goals set out in the Paris Agreement of limiting the increase in global average temperature to well below 2°C above preindustrial levels. On behalf of the emerging biofuel industry the message from the BfP is that although there are tremendous opportunities for advanced biofuels, there is a need for a stable regulatory framework in the form of a mandate for advanced biofuels.

The technology is there but it requires supportive legislation with a clear mandate in order the meet the high market demand for advanced biofuels globally.

The International Energy Agency (IEA, 2018) predicts that bioenergy could remain the most important renewable energy beyond the middle of this century if internationally agreed energy strategies are implemented in full. Furthermore, if the anticipated deployment of BECCS at scale is added to this there will be significant implications for future energy demand and supply. However, there is a great deal of questioning the ability to sustainably scale up feedstock, including concerns over indirect land use change. Furthermore, feedstock supply may be inconveniently located with respect to processing facilities. The result is likely to be that developed countries will increasingly import biomass from less developed countries. Clearly, national policies aimed at increasing availability of indigenous sources are needed, but also there is a need to source sustainable biomass from key regions around the world and develop global biomass trade markets (Gough et al., 2018).

Popp et al. (2014) suggest that a breakthrough in biomass demand could come as biomass becomes a mainstream commodity in commercial markets in standard form like pellets. Pellets may become a widespread commodity, efficiently transported internationally. The international bioenergy market is expected to have a wide range of suppliers from several regions of the world, so the importation of bioenergy will not be affected by the same geopolitical concerns as are oil and natural gas. It is therefore possible to envisage a growing trade in solid biomass (pellets and chips) and liquid biofuel with the adoption of sustainability criteria in the international market place. Standards and certificates will harmonise the markets, but they clearly should continuously change to take into account scientific advancements in the design and production of new materials and changing applications.

7 Policies and markets

The future for *Miscanthus*

Introduction

As we saw in the previous chapter, currently most of the advanced biofuel conversion technologies are commercially immature and the availability of suitable feedstocks is often limited. If feedstock production is to be scaled up to meet future commitments such as the European Union 2030 and 2050 climate objectives, policy makers will need to put the appropriate measures and incentives in place so that investors' confidence is boosted, particularly in the short term. This could allow the advanced biofuels sector to move beyond the current demonstration-scale plants towards increased plant sizes and scales of operation not observed before. However this needs to be matched by the incentivisation and mobilisation of the lignocellulosic feedstock sector to meet the demands of the energy suppliers.

There are a number of major issues on which there is still significant disagreement amongst policy makers. They include: (i) a lack of clarity on the methodologies used to calculate the mitigation potential of bioenergy, and in particular the calculation of carbon neutrality; (ii) the impact of large scale deployment of bioenergy on other biomass sectors; (iii) the comparative efficiency of bioenergy compared with other renewables; (iv) the suitability of different feedstocks to deliver short-term decarbonisation; and (v) the time horizons for the policy effects.

The main issues relating to market coordination and the effective operation of the market are the following:

- Agricultural producers need to massively invest in production of new high-yielding lignocellulosic crops and in industrialisation of the process of harvesting, conditioning and transport of the feedstock.
- Developers of the advanced conversion technologies, including innovators and industrial investors, need to invest in research and innovation to build conversion technologies of mass production and integrate the technologies into complex bio-refineries.
- Consumers, particularly in the transport sector, will need to build confidence in the new fuel substitutes. This will require education and a significant time period to develop.

Bioenergy currently constitutes a largely decentralised sector, composed mainly of small and medium sized companies, but this may change in the future should massive upscaling occur. Employment opportunities associated with bioenergy and biofuel developments are inherently local and are often based in rural areas, which currently experience high unemployment rates. The development of this sector clearly offers opportunities for increased employment in rural areas. For example, a US study of the impacts of advanced fuel production showed that about 65% of advanced biofuel jobs are related to feedstock production, supply harvesting and transport, and provide valued jobs and income creation in rural areas. Other high value jobs will also be created in areas of research and development, engineering, construction and processing operation.

The aim of this chapter is to review the possible future development of policies which should create markets for bioenergy crops, with a particular focus on *Miscanthus*.

Renewable energy policies

Essentially, all policies are directed at ensuring that they only support bioenergy use in the right circumstances. These can be summarised as follows:

- Policies to support bioenergy should deliver genuine carbon reductions that help meet national carbon emission objectives.
- Support for bioenergy should make a cost effective contribution to national carbon emission objectives in the context of the overall energy goals.
- Support for bioenergy should maximise the overall benefits and minimise costs across the national economy.
- When policies promote increased supply of bioenergy, this should only be done sustainably, i.e. in the light of assessments of impacts on food security and biodiversity.

Both the EU and the USA have been at the forefront of policy development in relation to use of biofuels. The energy crop markets operate within policy environments that are shaped by both energy policy and agricultural policy, and analysis shows that the interdependency between these policy areas determines the rate and level of adoption. In the EU, the Renewable Energy Directive (RED) (European Commission, 2009), established an overall policy for the production and promotion of energy from renewable sources. The expansion of biofuel production in the EU is largely policy driven and predicated on the role that biofuels can play in reducing GHG emissions and mitigating climate change (Gerssen-Gondelach et al., 2016). RED requires the EU to meet at least 20% of its total energy needs with renewables by 2020. All EU countries must also ensure that at least 10% of their transport fuels come from renewable sources by 2020. In November 2016 the European Commission published a proposal for a revised Renewable Energy Directive aimed at making the EU a global leader in renewable energy and ensuring that a target of at least 27%

renewables in the final energy consumption in the EU by 2030 is met. In June 2018 the European Parliament and the European Council reached a deal on a revised Renewable Energy Directive (REDII) that sets revised targets for renewables. For biofuels, the agreement states that at least 14% of transportation fuel must come from renewable sources by 2030. First-generation, crop-based biofuels are capped at 2020 levels. In addition, the share of advanced biofuels must be at least 1% in 2025 and at least 3.5% in 2030. Food crops, such as palm oil, that result in high indirect land use change (iLUC) are to be phased out through a certification process for low-iLUC biofuels. The GHG savings criteria for biofuels, biogas and bioliquids began at 50% before 2015, increased to 60% after 2015 and is 65% after 2021. For electricity, heating and cooling production from biomass fuels, the GHG reduction criteria are set at 70% after 2021 and 80% after 2026, where percentage reductions are compared to fossil fuels. Furthermore, biofuels cannot be grown in areas converted from land with previously high carbon stocks such as grasslands or forests. In addition, biofuels cannot be produced from raw materials obtained from land with high biodiversity such as primary forests or highly diverse grasslands. Only biofuels that comply with the criteria can receive government support or count towards national renewable energy targets. The Common Agricultural Policy (CAP) aimed at farmers in the EU, also has a role to play in influencing biomass production. The EU CAP regulations for 2014–2020 have three 'greening measures' directed to climate change mitigation and biodiversity conservation. Under the greening measures farmers receive a greening payment as a top-up to the basic payment scheme for practices that are beneficial for the climate and environment. Areas under energy crops such as willows and *Miscanthus* are currently eligible for the greening payment but uncertainty about the continuation of this measure post 2020 creates uncertainty for long-term establishment of energy crops.

In the USA liquid biofuel production has been particularly encouraged as a replacement for gasoline. The role of biofuels is to allow for energy independence and security as well as its contribution to the revitalisation of rural infrastructure and economy. Their role in combating climate change by reducing net CO_2 emissions is probably less appreciated but nevertheless makes a significant contribution. Currently, the large scale production of cellulosic biofuels is in the take-off stage. In the USA, POET-DSM Advanced Biofuels opened a commercial scale cellulosic ethanol plant in September 2014 and Abengoa opened a 25 million gallon cellulosic biofuel plant in Hugoton, Kansas. The introduction of the USA Renewable Fuels Standard 2 mandates for cellulosic biofuel, to be produced by 2022, has led to the commercialisation of several biofuel production processes in the USA (Searle and Malins, 2014).

The Energy Independence and Security Act of 2007, established in the USA the Renewable Fuel Standard (RFS), which sets the annual targets for biofuels from various feedstocks, including the amount of ethanol that must be produced from cellulosic versus grain biomass sources. Huiburg et al. (2016) integrated an ecosystem and economic model to assess the potential of US biofuel production to meet the RFS of 136 billion litres. In contrast to some assertions that biofuels rely on reduced food production to achieve the anticipated greenhouse gas benefits, this exercise showed that changing the mix of biofuel feedstocks from corn to high-yielding perennial crops such as *Miscanthus* could meet the RFS mandate without significantly reducing food production and with modest GHG savings. In order to induce bioenergy production in the USA the Biomass Crop Assistance Programme (BCAP) was established by the Food, Conservation, and Energy Act in 2008 (the 2008 Farm Bill) and reauthorized in the Agriculture Act of 2014 (the 2014 Farm Bill). Support provided by the BCAP included matching payments, establishment payments, assistance with the cost of producer cost of collection, storage and transport to a biomass conversion facility. Miao and Khanna (2017 a and b) in a detailed assessment of the effectiveness of the BCAP found that it is significantly influenced by farmers' risk, credit availability and biomass prices. When prices are low, most BCAP payments go to corn stover, and higher prices are required before *Miscanthus* is favoured by a higher yield and larger establishment costs.

Global biomass resources

At both the regional and global scale there is great uncertainty about the future demands on biomass production and whether production is sufficient to meet the demands of the policy targets outlined above. From a review of 90 studies, Slade et al. (2014) estimated the contributions to future energy supplies of biomass in different forms to be 22–1,272 EJ for energy crops, 10–66 EJ for agricultural residues, 3–35 EJ for forestry residues, 12–120 EJ for wastes, and 60–230 EJ for forestry. Estimates of biomass potential range so widely because they are calculated using different assumptions. Those at the high end basically test the boundaries of what might be physically possible, based on a physiological knowledge of plant productivity and a generous allocation of land to energy crop cultivation. Those that are more conservative tend to explore the boundaries of what might be socially acceptable or environmentally possible, and the most pessimistic estimates assume that there is limited land available for energy crops because of the increasing demand for land for food. Less pessimistic estimates assume that increasing food crop yields per unit land area keeps pace with population growth. Under these circumstances limited good quality agricultural land could remain available for energy crop production. The most optimistic estimates of energy production assume that increases in food-crop yields could significantly outpace demand for food, with the result that very large areas of agricultural land as well as grassland and marginal land could be made available for energy crops (Slade et al., 2014).

In addition to uncertainty about the area of land available for production of biomass, the productivity of that land is also very uncertain. In many assessments a simplistic assumption is made that yields of less than 5 Mg ha^{-1} y^{-1} will be achieved on marginal and degraded land, while yields close to 15 Mg ha^{-1} y^{-1} are ascribed to good quality land. Furthermore, most studies do not identify specific energy crop species and assume that the best adapted crop for each area and land type will be used. In a rather more nuanced approach, Searle and Malins (2015) suggest that, within their suitable climate ranges, yields on 'non-arable' land are currently 3–15 Mg ha^{-1} y^{-1} for *Miscanthus*, 3–10 Mg ha^{-1} y^{-1} for poplar and willow SRF (Short Rotation Forestry), and 5–15 Mg ha^{-1} y^{-1} for Eucalyptus SRF. Another reason for uncertainty about biomass crop yields is that plot trials frequently record higher yields than are achieved by the farmer in practice (see Chapter 3). As a result, policy makers and other non-specialists often quote unrealistic projections of energy crop yields.

Uncertainty about crop yields is the main reason for uncertainty about how much sustainable biomass we can count on to meet future targets. The question is 'Are current estimates overly optimistic?' and the answer is that we don't know. Uncertainty about biomass resources also raises issues at the national scale about securing a bioenergy future without imports (Welfle et al., 2014). In fact, most energy strategies of European states assume the use of non-EU sourced biomass to meet the forecast demand (Scarlat et al., 2015). The UK provides an example of a nation with strong bioenergy aspirations but uncertain biomass resource availability. Welfle et al. (2014) used a Biomass and Resource Model, which reflects the key biomass supply-chain dynamics and interaction that determine biomass resource availability in the UK, and show that up to 44% of energy demand could be met by 2050 within the UK and without impacting on the food system. More than half of this demand would be met by residues from agriculture, forestry and industry, but specifically grown biomass crops would supply up to 22% of demand.

A further uncertainty is that access to bioenergy may be even more restricted because the rapidly expanding bio-based economy is expected to be a worldwide development and consequentially a significant part of the globally available biomass potential will be diverted for local demands and will not be available for importers such as the EU. This is likely to put even more pressure on the demand for indigenous production on land, particularly in the EU, and one consequence would be demand for conversion of high carbon soils, such as grassland, to biomass production which will inevitably lead to increased GHG emissions.

Finally, international commitments under the IPCC Paris Agreement, to prevent warming greater than 2°C above preindustrial temperatures, have the potential to be extremely disruptive in terms of future requirements for biomass. As decarbonising the global economy, in an attempt to reduce greenhouse gas emissions, becomes more and more difficult to achieve, terrestrial carbon dioxide removal will become an increasingly important option. The removal of CO_2 from the atmosphere, referred to as negative carbon emission, is an important concept for climate change mitigation (EASAC, 2018). Through the process of modelling of GHG emissions for the rest of this century to achieve below 2°C warming, it has been shown that negative emissions should start within the next one to two decades and increase through the second half of the 21st Century. Several technologies, referred to as NETs (Negative Emission Technologies) have been identified to achieve negative emissions, but the most widely supported is referred to as bioenergy with carbon capture and storage (BECCS) (Vaughan et al., 2018) (see Chapter 6). This would require the large scale cultivation of bioenergy and capturing the carbon dioxide released upon combustion or anaerobic digestion for long-term storage in geologic formations. However, the amount of land required to meet the biomass demand is highly uncertain (Krause et al., 2018). Consequently, estimates of the potential of BECCS to remove significant amounts of carbon from the atmosphere are regarded as controversial and arguably unrealistic (Fuss et al., 2014). Furthermore, the time frame involved in planning and establishing a climate mitigation project of this nature is probably several decades, while action, to be effective, is required within the next decade.

Food versus fuel and the use of marginal land

Food versus fuel is the dilemma regarding the risk of diverting farmland previously used for food crops to biofuels' production to the detriment of the food supply and leading to higher food prices. The biofuel and food price debate involves wide-ranging views, and is a long-standing, controversial one in the literature. There is disagreement about the significance of the issue, what is causing it, and what can or should be done to remedy the situation. An economic assessment report published by the Organisation for Economic Development and Cooperation (OECD, 2008) found that 'the impact of current biofuel policies on world crop prices, largely through increased demand for cereals and vegetable oils, is significant but should not be overestimated. Current biofuel support measures alone are estimated to increase average wheat prices by about 5 percent, maize by around 7 percent and vegetable oil by about 19 percent over the next 10 years.'

In exploiting the benefits of bioenergy, probably the most important question is 'How can we use bioenergy without jeopardising food production?' One option is to derive energy only from the residues of food crop or farm and food waste and bioproducts. However, more relevant to the development of *Miscanthus* as a source of bioenergy is, 'Can we use dedicated perennial lignocellulosic energy crops like *Miscanthus* to reduce or eliminate direct competition with food production?' The question of whether bioenergy crops compete with food production for additional land depends mainly on two issues. The first is whether the required increase in food production can be achieved through sustainable intensification rather than through expanding use of agricultural land. The second is whether sustainable bioenergy demands can be met by crops grown on marginal land that is not used or required for food production (Valentine and Clifton-Brown, 2012).

Certainly, the scale of land use required to meet bioenergy targets has raised doubts among many commentators about the future for biofuels (Smith, 2010). The concern is that large scale development of biofuels will potentially have significant negative impacts on agriculture and the environment, while having a relatively small impact on the energy sector. Searchinger et al. (2008) has been one of the loudest voices arguing that growing bioenergy crops cannot lead to a net reduction in emissions of GHGs and that in fact GHG emissions associated with large scale biofuel production could be significantly higher than those of replaced fossil fuels once emissions associated with indirect land use change (iLUC) are factored in.

The use of less productive or 'marginal land' for energy crops has been suggested as a way of avoiding the land controversies created by their expansion. This is because, it is argued, marginal land does not compete with food production which preferentially uses better quality land when available. However the term 'marginal land' is notoriously imprecise as definitions vary between 'land unsuitable for food production' to 'land that is economically marginal' (Shortall, 2013). The use of these two definitions could result in very different estimates of marginal land in different parts of the world. It is therefore important not to raise unrealistic expectations about the role of marginal land in overcoming biofuel land-use controversies. Furthermore, one of the obvious consequences of growing crops on poor quality marginal land is that yields are likely to be significantly lower than on better quality land. However, this should incentivise the selection in breeding programmes of genotypes that are better adapted to maximising productivity on marginal land (see Chapter 5).

Miscanthus production and bioenergy policies

As a result of the concerted efforts by researchers over the last 25 years to elucidate the beneficial characteristics of *Miscanthus* as an energy crop the policy makers are increasingly developing proposals that incorporate the use of *Miscanthus* and other perennial rhizomatous grasses. There has been increasing policy support for second generation cellulosic biofuels in the USA due to the concerns about energy security, climate change and most importantly food versus fuel competition for land posed by the development of grain-based biofuels. However, most plans for cellulosic biofuel production appear to use corn stover as feedstock although low yields per hectare and negative impact of removal of stover on soil organic carbon are disincentives (Liska et al., 2014). The Herbaceous Energy Crops Research Programme (HECP) funded by the US Department of Energy (DOE) was established in 1984 in which 35 potential herbaceous crops, of which 18 were perennial rhizomatous grasses (PRGs), were assessed and it was concluded that switchgrass was the native PRG that showed the greatest potential. In 1991 the DOE's Bioenergy Feedstock Development Programme (BFDP), evolving from the HECP, placed the focus of research on switchgrass as the 'model' crop system.

More recently, Miao and Khanna (2017a) in the USA have assessed the costs of meeting the cellulosic biofuel mandate with a focus on *Miscanthus*. They highlight the high cost of establishment as a major difficulty with this crop as it typically has a one to three year establishment period during which the farmer incurs the fixed cost of the establishment period, while forgoing returns that could have been earned under conventional crops. Furthermore, with a lifespan of 20 years it needs a long-term commitment of the land to the crop. The decision to convert land from existing use to a perennial energy crop is complex as it depends on the riskiness of establishing alternative crops. The Biomass Crop Assessment Programme (BCAP) was established to incentivise the production of bioenergy and bioproducts, but its effectiveness is significantly affected by farmers' risks and time preferences, credit availability and biomass prices (Miao and Khanna, 2017b). In terms of incentivising development of perennial grasses like *Miscanthus*, the availability of establishment cost subsidies is predicted to be far more effective than the crop insurance which is currently availed of by more than 80% of eligible crop land in the USA.

In Europe, *Miscanthus* and willow are the second generation energy crops most frequently referred to by policy makers and there have been several assessments of the prospects for *Miscanthus* development. In the UK, Lovett et al. (2009) used GIS-based suitability and yield mapping to identify land suitable for *Miscanthus* production. They found that although the highest biomass yields co-located with the food producing area on high grade land, when such high grade land is excluded, the UK policy-driven proposal for planting on 350,000 ha required between 4% and 28% of lower grade land, depending on the region. They concluded that utilising this area is not likely to impact on UK food security.

In another analysis by Wang et al. (2014) for Great Britain, they found that a combination of *Miscanthus* and Short Rotation Forestry (SRF) could contribute 5% of the heat and electricity demand using about 5% of the potentially available land. Alexander et al. (2015) used a range of potential policy scenarios to evaluate the cost effectiveness of the market in providing a source of low carbon renewable electricity and to achieve carbon emissions abatement. This showed that, in the UK, energy crops, and in particular *Miscanthus*, have the potential to deliver emissions abatements up to 4% of total UK emissions, and renewable electricity of up to 8% of UK electricity, or 3% of the primary energy demand. However this requires significant financial support and, if this is not forthcoming, they suggest that domestically grown perennial energy crops will only play a niche role and supply less than 0.2% of the UK energy demand.

Persuading farmers to grow *Miscanthus* crops

Apart from high and sustainable yields, the main attributes of *Miscanthus* that make it attractive to farmers as an energy crop are the low requirements of fertilisers and pesticides for cultivation and the low maintenance costs once established, as the crop needs only to be harvested once a year. It also offers economic benefits in the form of new markets and higher or more stable income opportunities. In essence it is a low input/high output crop. Another advantage is that it can be grown on poorer quality land (loosely defined as marginal land) so that it does not compete with land used for higher value food crops. Furthermore, during the crop cycle it increases the soil quality in terms of carbon content by the process of carbon sequestration. There are, of course, financial and technical challenges for the farmer which include the high cost of establishing the *Miscanthus* from rhizome, the planting process, and in many cases patchy establishment which results in substantially lower yields than anticipated (Zimmerman et al., 2014).

Despite these advantageous characteristics of *Miscanthus,* and the introduction in recent years of a number of schemes to encourage farmers to grow energy crops, the established area has been very low and the evidence is that since 2010 its use for biomass energy has stagnated. In fact, it is remarkably difficult to obtain reliable statistics about the cultivation and production of *Miscanthus,* so the information on its scale of use are limited. In a review of Europe by Lewandowski (2016), it was estimated that approximately 20,000 ha of *Miscanthus* was being grown. This included about 10,000 ha in the UK, 4,000 ha in Germany, 4,000 ha in France, 500 ha in Switzerland and 500 ha in Poland. As to its use, in the UK it was mainly used for co-firing with coal for electricity generation and in Germany mainly for domestic heating, while in Denmark a very small area of production was used for thatching. The reasons for the low production in Europe and the obvious lack of interest from farmers in establishing the crop is generally seen to be the relatively high costs of production of *Miscanthus,* particularly the cost of establishment, combined with the lack of a stable market with a reasonably high value for the crop.

In an economic evaluation of *Miscanthus* production, Witzel and Finger (2016) reviewed 51 studies to identify the factors influencing the adoption of *Miscanthus.* The majority of the studies were published after 2009 and refer to cases in both Europe and N. America. They concluded that economic viability of *Miscanthus* depends largely on uncertain assumptions about yield, price, and life-span. It was suggested that the major barriers to adoption are lack of established markets, high establishment costs, and uncertainties associated with the need for long-term commitments by farmers. There is clearly a need to assess the risks and uncertainties for farmers choosing to cultivate *Miscanthus.* Although subsidies are crucial for supporting *Miscanthus* as a new crop, it was shown that these are very mixed across the countries where it is grown.

In order to supply guidance to farmers who might be interested in taking up cultivating *Misscanthus,* there have been a number of publications at the national level which lay out the current 'best practice guidelines' (e.g. DEFRA, 2001; USDA, 2011; Caslin et al., 2011). These guidelines are designed to introduce farmers to a new crop and they give advice on the most appropriate location, land preparation, planting techniques and crop management to grow *Miscanthus* as a crop destined for bioenergy use.

In order to better understand the reasons for the reluctance of farmers to take up the cultivation of *Miscanthus,* a number of surveys of farmers have been carried out in different countries. Glithero et al. (2013) carried out an on-farm survey with 244 English arable farmers and found only 1.2% were currently growing *Miscanthus* and that 81.6% would not now consider growing it. The main reasons given for farmers not growing *Miscanthus* were varied but included impacts on land quality, lack of appropriate machinery, commitment of land for a long period of time, time to financial return, and profitability. However, reasons cited by the small number of farmers willing to grow *Miscanthus* included land quality improvement, ease of crop management, commitment of land for a long period of time, and profitability. It was concluded that enhanced information exchange through extension agents, providing market security and considering land reversion grants post-production might be useful policy considerations in the future (Glithero et al., 2013).

In Ireland, a similar survey by Augustenborg et al. (2012) asked farmers their opinions about and preferences for energy crops. *Miscanthus* (48%) and perennial grasses (33%) were the most favoured but there was a very low interest in growing any energy crops because of a perceived lack of knowledge of the economic benefits and the availability of crops. The majority of farmers (84%) agreed that they would use their land to produce whatever products gave them the most consistent gross margins and that long-term contracts and government support were necessary before there would be large scale adoption of energy crops. In another survey of farmers' attitudes to energy crops in Sweden, Paulrud and Laitila (2010) gave a rather different picture. Here it was found that the visual impact on the landscape and the rotation period of the energy crop appear to have a significant impact on the utility derived from growing the crop. Other significant factors affecting the willingness to grow energy crops were the age of the farmer, the size of farm and the geographical area. This study confirmed that farmers' attitudes rather than strictly economic benefits should be considered in the policy discussions regarding future energy crop production.

In a further European assessment, Gillich et al. (2018) looked at the potential for the regional supply of *Miscanthus* in southwestern Germany. They concluded that it has a relatively small potential, particularly given the current wood chip price, and that a one-time subsidy will be required to ensure any uptake. These payments, if awarded selectively, could be used to create clusters of crop production. Such clusters would ideally be located close to processing plants which would save transportation costs. Also, improved exchange of information among colleagues would be encouraged so that farmers with specific problems or uncertainties could support each other. However Gillich et al. (2018) hold the view that, as long as the industrial use and demand for lignocellulose is not fully developed, *Miscanthus* should not be directly subsidised. A Catch 22 situation!

Moving across the Atlantic to the USA, Miguez et al. (2012), in a meta-analysis of the effects of management factors on the take-up of *Miscanthus* by farmers in Illinois, found that the rate of adoption was slowed at first by the requirement for a significant initial investment and a delayed economic return as farmers cannot afford this temporary economic burden. However, they also found that economic costs and benefits were not the whole explanation for the adoption decision process. Other factors included the technical complexity, perceived risk, and the compatibility with current operations. It was concluded that the farmers' perspectives and goals differ from researchers or policy makers so that information needs to be targeted to the producer's needs and concerns in relation to any proposed innovation. Furthermore, it was demonstrated that there are clear differences among information needs of farmers in each region of Illinois as well as preferred channels for providing this. Consequently, information campaigns need to address the regional needs and channel them through the preferred media.

To summarise, the greatest potential benefit for farmers growing *Miscanthus* are that (i) there are low labour and machinery inputs after establishment due to its perennial nature, (ii) it fits well to arable farming practices in that early spring harvests avoid clashes with most other field operations and there is little maintenance during the rest of the year, (iii) the combination of low annual costs and reliable annual yields results in a stable income from the more marginal areas of the farm, and (iv) on farm use of *Miscanthus* for energy production can dramatically reduce the energy costs on the farm by substituting for bought-in fuel. McCalmont et al. (2017a) also suggest that concentrating agronomic efforts and resources away from the least productive 10% of farmed land area to more productive land, while retaining low input, high output perennial energy crops should enable intensification and diversification. Under these circumstances *Miscanthus* could be used as a break crop which improves soil quality and soil carbon content.

Conclusions

Scenarios indicate that bioenergy will provide 25% of global primary energy supply by 2050 (IEA, 2018) and it is now increasingly clear that as demand for biomass from many different sectors continues to increase, in the short term there is probably a need to restrain demand. In deciding on how biomass is optimally used in the future the main considerations are likely to be the level of financial investment required, the social and behavioural changes necessary to establish its use, the knock-on price effects on different goods and services and the effects of other renewable energy developments.

A number of conditions will need to be met, which should ensure the rational use of available bioenergy (Vaughan et al., 2018). These include:

- Strong regulation and governance of bioenergy production. Particularly protection of forests, agricultural land and biodiversity.
- Development of a well-functioning large scale bioenergy market.
- Technical developments in energy crops, energy conversion technologies and carbon capture technologies.

Delivering this inspiring but challenging vision will require a coordinated international effort involving a wide number of stakeholders. It will be up to individual countries and stakeholders to explore the potential of possible actions and implement a strategy given their circumstances. It will also be important for governments at all levels, academia, industry and finance institutions to work together to develop a comprehensive suite of actions for consideration and to collectively pave the way to a lower carbon future. Increased priority should be given to low carbon sustainable bioeconomy projects as a key part of renewable energy, climate change mitigation and 'green' financing portfolios, greatly increasing available resources. It will be necessary to deploy loan guarantees and other financial instruments to facilitate development, production and market deployment of low carbon fuels and bio-based products by the research community. This can be achieved by supporting high quality research into new and/or improved bio-based processes and products and conversion and utilisation systems optimised for bioenergy. This research should provide high quality evidence and analysis relating to the sustainability of bioenergy and bioproducts so as to build public confidence and consensus and ultimately produce technical advice to support government design of public policies for the bioeconomy.

The large scale utilisation of biofuels will require international cooperation, regulations and certification mechanisms. In addition, sustainability criteria must be established regarding the use of land and the mitigation of environmental impacts caused by biofuel production. Bioenergy will be produced from dedicated biomass crops, agricultural residues and waste, and this will be a driver of transformation in the way we use our resources and land. The currently inefficiently used land, degraded land and residues will be utilised intensively and in so doing value will be added to agricultural economies. But policies are required to maximise the benefits and minimise the potential side effects. It is necessary to identify crops that will maximise productivity on marginal land by having maximum efficiency in their use of solar energy, be efficient in their uses of water and fertilisers, as well as being resilient to the oncoming climate change.

Fundamentally, the sustainable bioeconomy of the future will depend on the efficient supply of bioenergy from limited bio-resources. It is patently obvious that the current efficiency of biomass production is too low to be able to fully replace the current and future demand for fossil fuels (MacKay, 2008). The photon-to-fuel conversion efficiency (PFCF) of bioenergy crops can be expressed as the percentage of light energy hitting the plant converted to chemical energy by photosynthetic light reactions and via a number of processes ending up as a fuel (Aro, 2016). Zhu et al. (2008) calculate that the theoretical limit for the conversion of solar energy to biomass is 4.6% for C_3 photosynthesis at 30°C and present-day atmospheric CO_2 concentrations, and 6% for plants like *Miscanthus* with C_4 photosynthesis. The highest solar energy conversion efficiencies reported for C_3 and C_4 crops are 2.4% and 3.7% respectively, and the average conversion efficiencies of major crops are close to about 1% (Monteith, 1977). Zhu et al. (2008) have pointed to exciting prospects for genetic engineering of more efficient plants, but progress in enhancing photosynthesis, improved stress tolerance, and increased water use efficiency has been slower than expected in the past decades. In fact in terms of utilisation, and in particular the development of liquid biofuel production, a more significant breakthrough might be the discovery of naturally occurring microbes that can break down lignin to give easier access to cellulose. Another approach to improving biofuel production would be to change the composition of the plant cell wall. Genome-based new technologies could be used to improve cell wall digestibility and significantly increase the photon-to-fuel conversion efficiency.

As the world's population continues to grow over the next 40 years, agricultural production will have to increase by 60% and it is at this point that the increasing demand for suitable land for biomass production competes with the need for food production and leads to conflicts between land use for food and that used for producing bioenergy crops. These problems will be exacerbated by the change in land productivity caused by climate change (erosion, water stress, temperature stress, increasing soil salinity). Clearly, policies for promoting biomass as an alternative energy source will need to take these potential land use conflicts into account. This issue is exacerbated by the significant uncertainties with regard to reliably estimating the biomass production potential due to differences in approaches, assumptions, and aggregation levels that need to be addressed. Besides biomass availability, the future application of biomass for energy production is also determined by interactions with other sustainable, low carbon, energy options and their relative costs. Globally, in order to decarbonise our economies rapidly we need to speed up significantly the development of advanced biofuels capacity and bring commercial scale plants on line much more rapidly than at present. Meanwhile, high feedstock prices and poor margins continue to challenge biofuel producers.

However, the largest barrier to achieving these targets is the scale at which enterprises will need to operate. The main issues regarding the viability of bioenergy plants are in the development of reliable, integrated biomass supply chains from cultivation, harvesting, transport, and storage to conversion and by-product use. Secure, long-term supply of sustainable feedstock, often by local supply chains, is essential to the economic viability of bioenergy plants. For example, in order to supply a commercial scale second-generation bioethanol plant, based on an assumed annual dry matter production in energy plantations of 10 Mg ha^{-1} y^{-1} and an ethanol yield of 20–30%, requires a plantation area ranging between 250 km^2 and 600 km^2 or the area within a circle of 20 km to 27 km respectively. With these criteria for establishing a bioethanol plant, large areas of the more densely populated parts of the world are excluded, including much of Europe. Ultimately, the achievement of bioenergy targets will depend on both technological advances as well as non-technological factors, such as economies of scale, resulting from an increase in produced and installed capacity, risk-finance for first-of-a-kind manufacturing pilot lines, and demonstration of small, commercial-scale bioenergy, biofuels and biomass co-fired CHP plants, as well as standardisation. However, ligno-cellulosic bio-refineries have different economic and abatement characteristics from power plants. These differences will alter the energy crops market's potential for emission abatement and response to policy.

The responsibilities for policy development often lie in separate government departments making coordinated policy more difficult. A coherent and stable set of related policies is needed to ensure that the potential for energy crops to deliver significant emissions' abatement and to provide a source of renewable energy is achieved in a cost-effective manner. Farmers appear to be unwilling to grow crops without a more mature market, while potential investors are unwilling to develop the plants and technologies to create the demand and establish a market – a classic chicken and egg situation. Clearly the reasons for the lower than anticipated uptake to date needs to be understood, and potential stimuli to the market need to be identified. The energy crop market is a complex system involving human decision-making by many individuals working within an evolving policy market. The future of bioenergy, and the opportunities for energy crops such as *Miscanthus*, requires an appreciation of the complexity of the market and the willingness of players, along a complex chain from production to utilisation, to interact and collaborate to achieve their goals. *Miscanthus* for bioenergy is certainly not a silver bullet, but because of its almost unique characteristics of high productivity with low inputs it offers a significant opportunity to contribute to meeting one of our major challenges for the development of low carbon, renewable forms of energy.

Bibliography

Adams, P.W., Hammond, G.P. McManus, M.C. et al. (2011) Barriers to and drivers for UK bioenergy development, *Renewable and Sustainable Energy Reviews*, 15, 1217–1227.

Agostini, F., Gregory, A.S. and Richter, G.M. (2015) Carbon sequestration by perennial energy crops: Is the jury still out? *Bioenergy Research*, 8, 1057–1080.

Alexander, P., Moran, D. and Rounsevell, M.D.A. (2015) Evaluating potential policies for the UK perennial energy crop market to achieve carbon abatement and deliver a source of low carbon electricity, *Biomass and Bioenergy*, http://dx.doi.org/10.1016/l. biomass.2015.04.025.

Anderson, K. and Peters, G. (2016) The trouble with negative emissions, *Science*, 354, 182–183.

Anderson-Teixeira, K.J., Duvan, B.D., Long, S. et al. (2012) Biofuels on the landscape: Is "land sharing" preferable to "land sparing"? *Ecological Applications*, 22, 8, 2035–2048.

Arnoult, S. and Brancourt-Hulmel, M. (2015) A review on *Miscanthus* biomass production and composition for bioenergy use: Genotypic and environmental variability and implications for breeding, *Bioenergy Research*, 8, 502–526.

Aro, E.-M. (2016) From first generation biofuels to advanced solar biofuels, *Ambio*, 45 (Suppl. 1), S24–S23.

Arundale, R.A., Dohleman, F.G., Heaton, E.A. et al. (2014a) Yields of *Miscanthus x giganteus* and *Panicum virgatum* decline with stand age in the Midwestern USA, *GCB Bioenergy*, 6, 1–13.

Arundale, R.A., Dohleman, F.G., Voigt, T.B. et al. (2014b) Nitrogen fertilization does significantly increase yields of stands of *Miscanthus x giganteus* and *Panicum virgatum* in multiyear trials in Illinois, *Bioenergy Research*, 7, 408–416.

Atienza, S.G., Satovic, Z., Peterson, K.K. et al. (2003) Identification of QTLs influencing agronomic traits in *Miscanthus sinensis* Anderss. I. Total height, flag-leaf height and stem diameter, *Theoretical and Applied Genetics*, 107, 123–129.

Augustenborg, C.A., Finnan, J., McBennett, L. et al. (2012) Farmers' perspectives for the development of a bioenergy industry in Ireland, *GCB Bioenergy*, 4, 597–610.

Baldiini, M., da Borsa, F., Ferfuia, C. et al. (2017) Ensilage suitability and bio-methane yield of *Arundo donax* and *Miscanthus x giganteus*, *Industrial Crops and Products*, 95, 264–275.

Baxter, X.C., Darvell, L.I., Jones, J.M. et al. (2014) *Miscanthus* combustion properties and variations with *Miscanthus* agronomy, *Fuel*, 117, 851–869.

Beale, C.V. and Long, S.P. (1995) Can perennial C_4 grasses attain high efficiencies of radiant energy conversion in cool climates? *Plant Cell and Environment*, 18, 641–650.

Beale, C.V., Morison, J.I.L. and Long, S.P. (1999) Water use efficiency of C_4 perennial grasses in a temperate climate, *Agriculture and Forest Meteorology*, 96, 103–115.

Bessou, C., Ferchaud, F., Gabrielle, B. et al. (2011) Biofuels, greenhouse gases and climate change. A review, *Agronomy and Sustainable Development*, 31, 1–79.

Biofuture Platform (2018) Creating the biofuture: A report on the state of the low carbon bioeconomy. biofutureplatform.org.

Bok, J.P., Choi, H.S., Choi, J.W. and Choi, Y.S. (2013) Fast pyrolysis of *Miscanthus sinensis* in fluidized bed reactors: Characteristics of product yields and biocrude oil quality, *Energy*, 60, 44–52.

Bourke, D., Stanley, D., Emmerson, M. et al. (2014) Response of farmland biodiversity to the introduction of bioenergy crops: Effects of local factors and surrounding landscape context, *GCB Bioenergy*, 6, 275–289.

Bradshaw, J.D., Prasifka, J., Steffey, K.L. and Gray, M.E. (2010) First report of field populations of two potential aphid pests of the bioenergy crop *Miscanthus x giganteus*, *Florida Entomologist*, 93, 135–137.

Bridgewater, A.V. (2012) Review of fast pyrolysis of biomass and product upgrading, *Biomass and Bioenergy*, 38, 68–94.

Brock, A., Hoekman, S.K. and Unnasch, S. (2013) A review of variability in indirect land use change assessment and modelling in biofuel policy, *Environmental Science and Policy*, 29, 147–157.

Brosse, N., Dufour, A., Meng, X. et al. (2012) *Miscanthus*: A fast growing crop for biofuels and chemicals production, *Biofuels, Bioproducts & Biorefining*, 6, 580–592.

Bullard, M.J., Nixon, P.M.I. and Cheath, M. (1997) Quantifying the yield of *Miscanthus x giganteus* in the UK, *Aspects of Applied Biology*, 49, 199–206.

Burner, D.M., Hale, A.L., Carver, P. et al. (2015) Biomass yield comparisons of giant reed, and miscane grown under irrigated and rainfed conditions, *Industrial Crops and Products*, 76, 1025–1032.

Busby, P.E., Ridout, M. and Newcombe, G. (2016) Fungal endophytes: Modifiers of plant disease, *Plant Molecular Biology*, 90, 645–655.

Campbell, J.E., Lobel, D.B., Genova, R.C. and Field, C.R. (2008) The global potential of bioenergy on abandoned agricultural land, *Environmental Science and Technology*, 42, 5791–5794.

Cappelletto, P., Mongandini, F., Barberi, B. et al. (2000) Papermaking pulps from the fibrous fraction of *Miscanthus x giganteus*, *Industrial Crops and Products*, 11, 205–210.

Carroll, J. and Finnan, J. (2012) Physical and chemical properties of pellets from energy crops and cereal straws, *Biosystems Engineering*, 112, 151–159.

Casler, M.D. (2005) Ecotypic variation among Switchgrass populations from the Northern USA, *Crop Science*, 45, 388–398.

Caslin, B., Finnan, J. and Easson, L. (Eds.) (2011) *Miscanthus Best Practice Guidelines*. Teagasc & AFBI. Ireland.

Chaitanya, K.V., Rama Krishna, C., Venkata Ramana, G. and Khasim Beebi, S.K. (2014) Salinity stress and sustainable agriculture: A review, *Agricultural Reviews*, 35, 34–41.

Chen, C-L., van-der Schoot, H., Dehghan, S. et al. (2017) Genetic diversity of salt tolerance in Miscanthus. *Frontiers in Plant Science*, doi:10.3389/fpls.2017.00187.

Cherubini, F. and Stromman, A.H. (2011) Life cycle assessment of bioenergy systems: State of the art and future challenges, *Bioresource Technology*, 102, 437–451.

Christian, D.G., Lamptey, J.N.L., Forde, S.M.D. and Plumb, R.T. (1994) First report of barley yellow dwarf luteovirus on *Miscanthus* in the United Kingdom, *European Journal of Plant Pathology*, 100, 167–170.

Christian, D.G. and Haase, E. (2001) Agronomy of *Miscanthus*. In: Jones, M. and Walsh, M. (Eds.) *Miscanthus for Energy and Fibre*. James and James, London, pp. 21–45.

Christian, D.G., Riche, A.B. and Yates, N.E. (2008) Growth, yield and mineral content of *Miscanthus x giganteus* grown as a biofuel for 14 successive harvests, *Industrial Crops and Products*, 28, 320–327.

Clark, L.V., Stewart, J.R., Nishiwaki, A. et al. (2015) Genetic structure of *Miscanthus sinensis* and *Miscanthus sacchariflorus* in Japan indicates a gradient of bidirectional but asymmetric introgression, *Journal of Experimental Botany*, 66, 4213–4225.

Clark, L.V., Dzyubenko, E., Dzyubenko, N. et al. (2016) Ecological characteristics and in situ genetic association for yield-component traits of wild *Miscanthus* from eastern Russia, *Annals of Botany*, 118, 941–945.

Clifton-Brown, J.C. and Jones, M.B. (1997) The thermal response of leaf extension rate in genotypes of the C4 grass *Miscanthus*: An important factor in determining the potential productivity of different genotypes, *Journal of Experimental Botany*, 48, 1573–1581.

Clifton-Brown, J.C. and Lewandowski, I. (2000a) Overwintering problems of newly established *Miscanthus* plantations can be overcome by identifying genotypes with improved rhizome cold tolerance. *New Phytologist*, 148, 287–294.

Clifton-Brown, J.C. and Lewandowski, I. (2000b) Water use efficiency and biomass partitioning of three different *Miscanthus* genotypes with limited and unlimited water supply. *Annals of Botany*, 86, 191–200.

Clifton-Brown, J.C., Lewandowski, I., Andersson, B. et al. (2001) Performance of 15 *Miscanthus* genotypes at five sites in Europe. *Agronomy Journal*, 93, 1013–1019.

Clifton-Brown, J.C., Lewandowski, I., Bangerth, F. and Jones, M.B. (2002) Comparative responses to water stress in stay-green, rapid and slow senescing genotypes of the biomass crop, *Miscanthus*. *New Phytologist*, 154, 335–345.

Clifton-Brown, J.C., Neilson, B.M., Lewandowski, I. and Jones, M.B. (2000) The modelled productivity of *Miscanthus x giganteus* (GREEF et DEU) in Ireland. *Industrial Crops and Products*, 12, 97–109.

Clifton-Brown, J., Chiang, Y-C., Hodkinson, T.R. (2008) *Miscanthus*: Genetic resources and breeding potential to enhance bioenergy production. In: Vermerris, W. (Ed.) *Genetic Improvement of Bioenergy Crops*. Springer, New York, pp. 273–294.

Clifton-Brown, J., Hastings, A., Mos, M. et al. (2017a) Progress in upscaling *Miscanthus* biomass production for the European bio-economy with seed-based hybrids, *GCB Bioenergy*, 9, 6–17.

Clifton-Brown, J., McCalmont, J. and Hastings, A. (2017b) Development of *Miscanthus* as a bioenergy crop. In: Love, J. and Bryant, J.A. (Eds.) *BioFuels and Bioenergy*, First Edition. John Wiley & Sons, pp. 119–131.

Clifton-Brown, J., Harfouche, A., Casler, M.D. et al. (2018) Breeding progress and preparedness for mass-scale deployment of perennial lignocellulosic biomass crops switchgrass, *Miscanthus*, willow and poplar. *GCB Bioenergy*, doi:10.1111/gcbb.12566.

Coelho, S.T. et al. (2012) Land and water: Linkages to bioenergy, Chapter 20. In: *Global Energy Assessment – Towards a Sustainable Future*. Cambridge University Press, Cambridge and New York and the International Institute for Applied Systems Analysis, Laxenburg, Austria, pp. 1459–1562.

Cosentino, S.L., Patane, C., Sanzone, E. et al. (2007) Effects of soil water content and nitrogen supply on the productivity of *Miscanthus x giganteus* Greef et Deu. in a Mediterranean environment, *Industrial Crops and Products*, 25, 75–88.

Coyle, W. (2007) The future of biofuels: A global perspective. *Amber Waves*, November, 24–29. www.ers.usda.gov/amberwaves.

Creutzig, F., Ravindranath, N.H., Berndes, G. et al. (2015) Bioenergy and climate change mitigation: an assessment, *GCB Bioenergy*, 7, 916–944.

Da Costa, R.M.F., Pattathil, S., Avci, U. et al. (2017) A cell wall reference profile for *Miscanthus* bioenergy crops highlights compositional and structural variations associated with development and organ origin, *New Phytologist*, 213, 1710–1725.

de Leon, N. and Coors, J.G. (2008) Genetic improvement of corn for lignocellulosic feedstock. In: Vermerris, W (Ed.) *Genetic Improvement of Bioenergy Crops*. Springer, New York, pp. 185–210.

Daly, C., Halbleib, M.D., Hannaway, D.B. et al. (2018) Environmental limitation mapping of potential biomass resources across the conterminous United States, *GCB Bioenergy*, doi:10.1111/gcbb.12496.

Dauber, J., Cass, S., Gabriel, D., Harte, K. et al. (2014) Assessing the impact of within crop heterogeneity in young *Miscanthus x giganteus* fields on economic feasibility and soil carbon sequestration, *GCB Bioenergy*, 6, 566–576, 455–467.

Dauber, J. and Bolte, A. (2014) Bioenergy: Challenge or support for the conservation of biodiversity? *GCB Bioenergy*, 6, 180–182.

Dauber, J., Cass, S., Gabriel, D et al. (2015) Yield-biodiversity trade-off in patchy fields of *Miscanthus x giganteus*, *GCB Bioenergy*, 7, 455–467.

Davey, C.L., Jones, L.E., Squance, M. et al. (2017) Radiation capture and conversion efficiencies of *Miscanthus sacchariflorus, M. sinensis* and their naturally occurring hybrid *M.x giganteus*, *GCB Bioenergy*, 9, 385–399.

Davis, S.C., Anderson-Teixeira, K.J. and DeLucia, E.H. (2008) Life-cycle analysis and the ecology of biofuels, *Trends in Plant Science*, 14, 140–146.

Davis, S.C., Parton, W.J., Dohleman, F.G. et al. (2010) Comparative biogeochemical cycles of bioenergy crops reveal nitrogen-fixation and low greenhouse gas emissions in a *Miscanthus x giganteus* agro-ecosystem, *Ecosystems*, 13, 144–156.

DEFRA (2001) *Planting and Growing Miscanthus*. DEFRA Publications, London.

Dhugga, K.S. (2007) Maize biomass yield and composition for biofuels. *Crop Science*, 47, 2211–2227.

Djomo, S.N. and Ceulemans, R. (2012) A comparative analysis of the carbon intensity of biofuels caused by land use change, *GCB Bioenergy*, 4, 392–407.

Don, A., Osborne, B., Hastings, A. et al. (2012) Land-use change to bioenergy production in Europe: Implications for the greenhouse gas balance and soil carbon, *GCB Bioenergy*, 4, 372–391.

Dondini, M., Hastings, A., Gustavo, S. et al. (2009) The potential of *Miscanthus* to sequester carbon soils: Comparing field measurements in Carlow, Ireland to model predictions, *GCB Bioenergy*, 1, 413–425.

Dong, H., Liu, S., Clark, L. et al. (2018) Genetic mapping of biomass yield in three interconnected *Miscanthus* populations, *GCB Bioenergy*, 10, 165–185.

Donnison, I.S. and Fraser, M.D. (2016) Diversification and use of bioenergy to maintain future grassland, *Food Energy Security*, 5(2), 67–75.

Dornburg, V., van Vuuren, D., van der Ven, G. et al. (2010) Bioenergy revisited: Key factors on global potentials for bioenergy, *Energy and Environmental Science*, 3, 258–267.

Drewer, J., Finch, J.W., Lloyd, C.R. et al. (2012) How do soil emissions of N_2O, CH_4 and CO_2 from bioenergy crops differ from arable annual crops? *GCB Bioenergy*, 4, 408–419.

EASAC (2012) The current status of biofuels in the European Union, their environmental impacts and future prospects, *EASAC Policy Report 19*, www.easac.eu.

EASAC (2017) Multi-functionality and sustainability in the European Union's forests, *EASAC Policy Report 32*, www.easac.eu.

EASAC (2018) Negative emission technologies: What role in meeting Paris Agreement targets? *EASAC Policy Report 35*, www.easac.eu.

EISA (2007) Energy Independence and Security Act of 2007. *Public Law* 110–140, 110th Congress, EISA.

Emmerling, C. and Pude, R. (2016) Introducing *Miscanthus* to the greening measures of the EU Common Agricultural Policy, *GCB Bioenergy*, doi:10.1111/gcbb 12409.

European Commission (2009) Directive 2009/28/EC of the European Parliament and of the Council of 23 April 2009 on the promotion of the use of energy from renewable sources and amending and subsequently repealing Directives 2001/77/EC and 2003/30/EC. European Commission, Brussels, Belgium.

European Commission (2011) COM (2011) 885/final. Energy Roadmap 2050. European Commission, Brussels, Belgium.

European Commission (2013) COM (2013) 175 Report from the Commission to the European Parliament, the Council, the European Economic and Social Committee and the Committee of the Regions. Renewable energy progress report. European Commission, Brussels, Belgium.

European Community (2016) SET-Plan – Declaration of Intent on "Strategic targets for bioenergy and renewable fuels needed for sustainable transport solutions in the context of an Initiative for Global Leadership in Bioenergy" https://setis.ec.europa.eu/... setplan/declaration_action8_renewablefuels_bioenergy.pdf.

European Commission (2017a) Research and innovation perspective of the mid- and long-term potential for advanced biofuels in Europe, doi:10.2777/419893.

European Commission (2017b) Biomass supply and demand for a sustainable bioeconomy-exploring assumptions behind estimates, doi:10.2777/39314, European Commission, Brussels, Belgium.

European Commission (2018) COM (2018) 773 final. A clean planet for all. European Commission, Brussels, Belgium.

Faaij, A.P.C. (2006) Bio-energy in Europe: Changing technology choices, *Energy Policy*, 34, 322–342.

Farrell, A.D., Clifton-Brown, J.C., Lewandowski, I. et. al. (2006) Genotypic variation in cold tolerance influences the yield of *Miscanthus*, *Annals of Applied Biology*, 149, 337–345.

Fonteyne, S., Muylle, H., Lootens, P. et al. (2018) Physiological basis of chilling tolerance and early-season growth in *Miscanthus*, *Annals of Botany*, 121, 281–295.

Frydendal-Nielsen, S., Jorgensen, U., Hjorth, M. et al. (2017) Comparing methods for measuring the digestibility of *Miscanthus* in bioethanol or biogas processing, *GCB Bioenergy*, 9, 168–175.

Fuss, S., Canadell, J.G., Peters, G.P. et al. (2014) Betting on negative emissions. *Nature Climate Change*, 4, 850–853.

Gabrielle, B., Bamiere, L., Caldes, N. et al. (2014) Paving the way for sustainable bioenergy in Europe: Technical options and research avenues for large-scale biomass feedstock, *Renewable and Sustainable Energy Reviews*, 33, 11–25.

Gasparatos, A., Stromberg, P. and Takeuchi, K. (2011) Biofuels, ecosystem services and human wellbeing: Putting biofuels in the ecosystem services narrative, *Agriculture, Ecosystems and Environment*, 142, 111–128.

Gauder, M., Graeff-Honninger, S., Lewandowski, I. et al. (2012) Long-term yield and performance of 15 different *Miscanthus* genotypes in Southwest Germany, *Annals of Applied Biology*, 160, 126–136.

GBEP (2011) The Global Bioenergy Partnership, Sustainability Indicators for Bioenergy. www.globalbioenergy.org.

Gelfand, I., Sahajpal, R., Zhang, X. et al. (2013) Sustainable bioenergy production from marginal lands in the US Midwest, *Nature*, 493, 514–517.

Gerssen-Gondelach, S.J., Wicke, B., Borzecka, M. et al. (2016) Bioethanol potential from *Miscanthus* with low ILUC risk in the province of Lublin, Poland, *GCB Bioenergy*, 8, 909–924.

Gibbs, H.K., Johnson, M., Foley, J.A. et al. (2008) Carbon payback times for crop-based biofuel expansion in the tropics: The effects of changing yield and technology, *Environmental Research Letters*, 3, 1–10.

Gifford, J.M., Chai, W.B., Swaminathan, K. et al. (2015) Mapping the genome of *Miscanthus sinensis* for QTL associated with biomass productivity, *GCB Bioenergy*, 7, 797–810.

Gillich, C., Narjes, M., Krimley, T. and Lippert, C. (2018) Combining choice modelling estimates and stochastic simulations to assess the potential of new crops – The case of lignocellulosic perennials in southwestern Germany, *GCB Bioenergy*, doi:10.1111/gcbb.12550.

Glowaka, K., Jorgesen, U., Kjeldsen, J. B. et al. (2015) Can the exceptional chilling tolerance of C_4 photosynthesis found in *Miscanthus x giganteus* be exceeded? Screening of a novel *Miscanthus* germplasm collection, *Annals of Botany*, 115, 981–990.

Glowaka, K., Ahmed, A., Sharma, S. et al. (2016) Can chilling tolerance of C_4 photosynthesis in *Miscanthus* be transferred to sugarcane? *GCB Bioenergy*, 8, 407–418.

Glithero, N.J., Wilson, P. and Ramsden, S.J. (2013) Prospects for arable farm uptake of Short Rotation Coppice willow and miscanthus in England, *Applied Energy*, 107, 209–218.

Gnansounon, E., Dauriat, A., Villegas, J. and Panichheli, L. (2009) Life cycle assessment of biofuels. Energy and greenhouse gas balances, *Bioresource Technology*, 100, 4919–4930.

Gopalakrishnan, G., Cristina, M. and Salas, W. (2012) Modelling biogeochemical impacts of bioenergy buffers with perennial grasses for a row-crop field in Illinois, *GCB Bioenergy*, 4, 739–750.

Gough, C., Garcia-Freites, S., Jones, C. et al. (2018) Challenges to the use of BECCS as a keystone technology in pursuit of 1.5°C, *Global Sustainability*, 1, e5, 1–9.

Gough, C., Mander, S., Thornley, P. et al. (Eds.) (2018) *Biomass Energy with Carbon Capture and Storage (BECCS): Unlocking Negative Emissions*. John Wiley & Sons Ltd.

Hansson, J., Berndes, G., Englund, O. et al. (2018) How is biodiversity protection influenecing the potential for bioenergy feedstock production on grassland? *GCB Bioenergy*, doi:10.1111/gcbb.12568.

Hastings, A., Clifton-Brown, J., Wattenbach, M. et al. (2008) Potential of *Miscanthus* grasses to provide energy and hence reduce greenhouse gas emissions, *Agronomy for Sustainable Development*, 28, 465–472.

Hastings, A., Clifton-Brown, J., Wattenbach, M., Mitchell, C.P. and Smith, P. (2009a) The development of MISCANFOR, a new *Miscanthus* growth Model: Towards more robust yield predictions under different climatic and soil conditions, *GCB Bioenergy*, 1, 154–170.

Hastings, A, Clifton-Brown, J., Wattenbach, M. et al. (2009b) Future energy potential of *Miscanthus* in Europe, *GCB Bioenergy*, 1, 180–196.

Hastings, A., Mos, M., Yesufu, J.A., et al. (2017) Economic and environmental assessment of seed and rhizome propagated *Miscanthus* in the UK, *Frontiers in Plant Science*, doi:10.3389/fpls.2017.01058.

Haughton, A.J., Bohan, D.A., Clark, S.J. et al. (2016) Dedicated biomass crops can enhance biodiversity in the arable landscape, *GCB Bioenergy*, 8, 1071–1081.

Heaton, E.A., Voigt, T. and Long, S.P. (2004a) A quantitative review comparing the yields of two candidate C_4 perennial biomass crops in relation to nitrogen, temperature and water, *Biomass and Bioenergy*, 27, 21–30.

Heaton, E.A., Clifton-Brown, J., Voigt, T., et al. (2004b) *Miscanthus* for renewable energy generation: European Union experience and projections for Illinois, *Mitigation and Adaptation Strategies for Global Change*, 9, 433–451.

Heaton, E.A., Dohleman, F.G. and Long, S.P. (2008a) Meeting US biofuel goals with less land: The potential of *Miscanthus*, *Global Change Biology*, 14, 200–214.

Heaton, E.A., Flavell, R.B., Mascia, P.N. et al. (2008b) Herbaceous energy crop development: Recent progress and future prospects, *Current Opinion in Biotechnology*, 19, 202–209.

Helm, D. (2015) *Natural Capital – Valuing the Planet*. Yale University Press, New Haven and London.

Hodkinson, T.R., Renvoize, S.A. and Chase, M.W. (1997) Systematics in *Miscanthus*, *Aspects of Applied Biology*, 49, 189–198.

Hodkinson, T.R. and Renvoize, S. (2001) Nomenclature of *Miscanthus x giganteus* (Poaceae), *Kew Bulletin*, 56, 759–760.

Hodkinson, T.R., Chase, M.W., Lledo, M.D., et al. (2002a) Phylogenetics of *Miscanthus*, *Saccharum* and related genera (Saccharinae, Andropogoneae, Poaceae) based on DNA sequences from ITS nuclear ribosomal DNA and plastid trnL and trnL-F intergenic spacers, *Journal of Plant Research*, 115, 381–392.

Hodkinson, T.R., Chase, M.W. and Renvoize, S.A. (2002b) Characterization of a genetic resource collection for *Miscanthus* (Saccharubae, Andropogoneae, Poaceae) using AFLP and ISSR PCR, *Annals of Botany*, 89, 627–636.

Hodkinson, T.R., Klaas, M., Jones, M.B. et al. (2015) *Miscanthus* a case study for the utilization of natural genetic variation, *Plant Genetic Resources: Characterization and Utilization*, 13, 219–237.

Holder, A.J., McCalmont, J.P., Rowe, R. et al. (2018) Soil N_2O emissions with different reduced tillage methods during the establishment of *Miscanthus* in temperate grassland, *GCB Bioenergy*, 11, 539–549, doi:10.1111/gcbb.12570.

Holland, R.A., Eigenbrod, F., Muggeridge, A. et al. (2015) A synthesis of the ecosystem services impact of second generation bioenergy crop production, *Renewable and Sustainable Energy Reviews*, 46, 30–40.

Houben, D., Sonnet, P. and Cornelis, J-T. (2013) Biochar from *Miscanthus*: A potential silicon fertilizer, *Plant and Soil*, doi:10.1007/s11104–11013–1885–1888.

Houghton, J.H. (2004) *Global Warming: The Complete Briefing*. CUP, Cambridge.

Hughes, J.K., Lloyd, A.J., Huntingford, C. et al. (2010) The impact of extensive planting of *Miscanthus* as an energy crop on future CO_2 atmospheric concentrations, *GCB Bioenergy*, 2, 79–88.

Huiburg, T.W., Wang, W-W., Khanna, M. et al. (2016) Impacts of a 32-billion-gallon bioenergy landscape on land and fossil fuel use in the US, *Nature Energy*, doi:10.1038/NENERGY.2015.5.

Hwang, O-J., Cho, M-A., Han, Y-J. et al. (2014) Agrobacterium-mediated genetic transformation of *Miscanthus sinensis*, *Plant Cell Tissue and Organ Culture*, 117, 51–63.

IEA (2011) *Technology Roadmaps – Biofuels for Transport*. OECD/IEA, Paris.

IEA (2013) *World Energy Outlook*. OECD/IEA, Paris.

IEA (2018) *Renewables 2018 – Analysis and Forecasts to 2023*. OECD/IEA, Paris.

IPCC (2011) *Special Report on Renewable Energy Sources and Climate Change Mitigation*. (ed. Change WGI-MOC).

Jensen, E., Farrar, K., Thomas-Jones, S. et al. (2011) Characterisation of flowering diversity in *Miscanthus* species, *GCB Bioenergy*, 3, 387–400.

Jensen, E., Robson, P., Farrrar, K. et al. (2016) Towards *Miscanthus* combustion quality improvement: The role of flowering and senescence, *GCB Bioenergy*, doi:10.1111/gebb.12391.

Jiao, X., Korup, K., Neumann Andersen, M., Sacks, E.J. et al. (2016) Can *Miscanthus* C_4 photosynthesis compete with *Festulolium* C_3 photosynthesis in a temperate climate? *GCB Bioenergy*, 9, 18–30.

Jones, H.G. and Jones, M.B. (1989) Introduction: Some terminology and common mechanisms. In: Jones, H.G., Flowers, T.J and Jones, M.B. (Eds.) *Plants Under Stress*. SEB Seminar Series, 39, CUP, Cambridge.

Jones, M.B. and Walsh, M. (Eds.) (2001) *Miscanthus for Energy and Fibre*. James and James, London.

Jones, M.B. and Donnelly, A. (2004) Carbon sequestration in temperate grassland ecosystems and the influence of management, climate and elevated CO_2, *New Phytologist*, 164, 423–429.

Jones, M.B. (2011) C4 species as energy crops. In: Raghavendra, A. and Sage, R.F. (Eds.) *C_4 Photosynthesis and Related CO_2 Concentrating Mechanisms*, Springer, The Netherlands, pp. 379–397.

Jones, M.B., Finnan, J. and Hodkinson, T.R. (2014) Morphological and physiological traits for higher biomass production in perennial rhizomatous grasses grown on marginal land, *GCB Bioenergy*, doi:10.1111/gebb.12203.

Jones, M.B., Zimmerman, J. and Clifton-Brown, J. (2016) Long term yields and soil carbon sequestration from *Miscanthus* – a review. In: Barth, S., Murphy-Bokern, D., Kalinina, O., Taylor, G. and Jones, M. (Eds.), *Perennial Biomass Crops for a Resource-Constrained World*. Switzerland: Springer International Publishing, pp. 43–49.

Jorgesen, U. (2011) Benefits versus risks of growing biofuel crops: The case of *Miscanthus*. *Current Opinions in Environmental Sustainability*, 3, 24–30.

Kalinina, O., Nunn, C., Sanderson, R. et al. (2017) Extending *Miscanthus* cultivation with novel germplasm at six contrasting sites, *Frontiers in Plant Science*, 8.

Kandel, T.P., Hastings, A., Jorgensen, U. et al. (2016) Simulation of biomass yield of regular and chilling tolerant *Miscanthus* cultivars and reed canary grass in different climates of Europe, *Industrial Crops and Products*, 86, 329–333.

Keymar, D.P. and Kent, A.D. (2014) Contribution of nitrogen fixation to first year *Miscanthus x giganteus*, *GCB Bioenergy*, 6, 577–586.

Kiesel, A. and Lewandowski, I. (2017) *Miscanthus* as a biogas – cutting tolerance and potential for anaerobic digestion, *GCB Bioenergy*, 9, 153–167.

Kim, K.W. (2015) Three-dimensional surface recostruction and in situ site-specific cutting of the teliospores of *Puccinia miscanthi* causing leaf rust of the biomass plant *Miscanthus sinensis*, *Micron*, 73, 15–20.

Kluts, I., Wicke, B., Leemans, R. and Faaij, A. (2017) Sustainability constraints in determining European bioenergy potential: A review of existing studies and steps forward, *Renewable and Sustainable Energy Reviews*, 69, 719–734.

Koh, L.P. and Ghazoul, J. (2008) Biofuels, biodiversity and people: Understanding the conflicts and finding opportunities. *Biological Conservation*, 141, 2246–2450.

Krause, A., Li, W. and Muller, C. (2018) Large uncertainty in carbon uptake potential of land-based climate-change mitigation efforts, *Global Change Biology*, doi:10.1111/gcb.14144.

Krol, D.J., Jones, M.B., Williams, M. et al. (2019) The effects of land use change from grassland to bioenergy crops *Miscanthus* and reed canary grass on nitrous oxide emissions, *Biomass and Bioenergy*, 120, 396–403.

Larsen, S.U., Jorgensen, U., Kjeldsen, J.B. et al. (2014) Long term *Miscanthus* yields by location genotype, row distance, fertilization and harvest season, *Bioenergy Research*, 7, 620–635.

Lask, J., Wagner, M., Trindadc, L. and Lewendowski, I. (2018) Life cycle assessment of ethanol production from *Miscanthus*: A comparison of production pathways at two European sites, *GCB Bioenergy*, doi:10.1111/gcbb.12551.

Lee, D.K., Aberle, E., Anderson, E.K. et al. (2018) Biomass production of herbaceous energy crops in the United States: Field trial results and yield potential maps from the multiyear regional feedstock partnership, *GCB Bioenergy*, doi:10.1111/gebb.12493.

Lee, M-S, Wycislo, A., Guo, J. et al. (2017) Nitrogen fertilization effects on biomass production and yield compoents of *Miscanthus x giganteus*, *Frontiers in Plant Science*, doi:10.3389/fpls.2017.00544.

Lee, W-C. and Kuan, W-C. (2015) *Miscanthus* as cellulosic biomass for bioethanol production, *Biotechnology Journal*, 10, 840–854.

Lesur, C., Jeuffroy, M-H., Makowski, D. et al. (2013) Modeling long-term yield trends of *Miscanthus x giganteus* using experimental data across Europe, *Field Crops Research*, 140, 252–260.

Lewandowski, I. (2016) The role of perennial biomass crops in a growing bioeconomy. In: S. Barth et al., *Perennial Biomass Crops for a Resource-Constrained World*. Springer International Publishing, Switzerland.

Lewandowski, I. and Kicherer, A. (1997) Combustion quality of biomass: Practical relevance and experiments to modify the biomass quality of *Miscanthus x giganteus*, *European Journal of Agronomy*, 6, 163–177.

Lewandowski, I. and Heinz, A. (2003) Delayed harvest of *Miscanthus* – influences on biomass quantity and quality and environmental impacts of energy production, *European Journal of Agronomy*, 19, 45–63.

Lewandowski, I. and Schmidt, U. (2006) Nitrogen, energy and land use efficiencies of *Miscanthus*, reed canary grass and triticale as determined by the boundary line approach, *Agriculture, Ecosystems and Environment*, 11, 335–346.

Lewandowski, I., Clifton-Brown, J.C., Scurlock, J.M.O. et al. (2000) *Miscanthus*: European experience with a novel energy crop, *Biomass and Bioenergy*, 19, 209–227.

Lewandowski, I., Clifton-Brown, J.C., Andersson, G. et al. (2003a) Environment and harvest time affects the combustion qualities of *Miscanthus* genotypes, *Agronomy Journal*, 95, 1274–1280.

Lewandowski, I., Scurlock, J.M.O., Lindvall, E. et al. (2003b) The development and current status of perennial rhizomatous grasses as energy crops in the US and Europe, *Biomass and Bioenergy*, 25, 335–361.

Liska, A.J., Yang, H., Milner, M. et al. (2014) Biofuels from crop residue can reduce soil carbon and increase CO_2 emissions, *Nature, Climate Change*, 4, 398–401.

Long, S.P. (1983) C4 photosynthesis at low temperatures, *Plant Cell and Environment*, 6, 345–363.

Long, S.P. (1999) Environmental responses. In: Sage, R.F. and Monson, R.K. (Eds.) *C4 Plant Biology*, Academic Press, San Diego, CA, pp. 215–249.

Long, S.P., Zhu, X-G., Naidu, S.L. et al. (2006) Can improvement in photosynthesis increase crop yields? *Plant, Cell and Environment*, 29, 315–330.

Long, S. and Spence, A.K. (2013) Towards cool C4 crops, *Annual Review of Plant Biology*, 64, 701–722.

Lovett, A.A., Sunnenberg, G.M., Richter, G.M. et al. (2009) Land use implications of increased biomass production identified by GIS-based suitability and yield mapping for *Miscanthus* in England, *Bioenergy Research*, 2, 17–28.

Lovett, A., Sunnenberg, G. and Dockerty, T. (2014) The availability of land for perennial energy crops in Great Britain. *GCB Bioenergy*, 6, 99–107.

MacKay, DJC. (2008) *Sustainable Energy – without the hot air*. UIT, Cambridge.

Malinowska, M., Donnison, I.S., Robson, P.R.H. (2017) Phenomics analysis of drought responses in *Miscanthus* collected from different geographical locations, *GCB Bioenergy*, doi:10.1111/gcbb.12350.

Mangold, A., Lewandowski, I., Mohring, J. et al. (2018) Harvest date and leaf: Stem ratio determines methane hectare yield of *Miscanthus* biomass, *GCB Bioenergy*, doi:10.1111/gcbb.12549.

Mathew, I., Shimelis, H., Mutema, M. et al. (2017) What crop type for atmospheric carbon sequestration: results from a global data analysis, *Agriculture, Ecosystems and Environment*, 243, 34–46.

Matlaga, D.P. and Davis, A.S. (2013) Minimizing invasive potential of *Miscanthus x giganteus* grown for bioenergy: Identifying demographic thresholds for population growth and spread, *Journal of Applied Ecology*, 50, 479–487.

Matlaga, D.P., Schutte, B.J., Davis, A.S. (2012) Age-dependent demographic rates of the bioenergy crop *Miscanthus x giganteus* in Illinois, *Invasive Plant Science and Management*, 5, 238–248.

McCalmont, J.P., Hastings, A., McNamara, N.P. et al. (2017a) Environmental costs and benefits of growing *Miscanthus* for bioenergy in the UK, *GCB Bioenergy*, doi:10.1111/gcbb,12294.

McCalmont, J.P., McNamara, N.P., Donnison, I.S. et al. (2017b) An interyear comparison of CO_2 flux and carbon budget at a commercial-scale land-use transition from semi-improved grassland to *Miscanthus x giganteus*, *GCB Bioenergy*, 9, 229–245.

McLaughlin, S.B. and Kszos, L.A. (2005) Development of switchgrass (*Panicum virgatum*) as a bioenergy feedstock in the United States, *Biomass and Bioenergy*, 28, 515–535.

Meehan, P., Finnan, J.M. and McDonnell, K.P. (2013) The effect of harvest date and harvest method on the combustion characteristics of *Miscanthus x giganteus*, *GCB Bioenergy*, 5, 487–496.

Menon, V. and Rao, M. (2012) Trends in bioconversion of lignocellulose: Biofuels, platform chemicals and bio-refinery concept, *Progress in Energy and Consumption*, 38, 522–550.

Miao, R. and Khanna, M. (2017a) Costs of meeting cellulosic biofuel mandate with perennial energy crops: Implications for policy, *Energy Economics*, 64, 321–334.

Miao, R. and Khanna, M. (2017b) Effectiveness of the biomass crop assistance program: Roles of behavioural factors, credit constraint, and program design, *Applied Economic Perspectives and Policy*, 39(4), 584–608.

Miguez, F.E., Zhu, X., Humphries, S. et al. (2009) A semimechanistic model predicting the growth and production of the bioenergy crop *Miscanthus giganteus*: Description, parameterization and validation, *GCB Bioenergy*, 1, 282–296.

Miguez, F.E., Maughan, M., Bollero, G.A. et al. (2012) Modeling spatial and dynamic variation in growth, yield, and yield stability of the bioenergy crops *Miscanthus x giganteus* and *Panicum virgatum* across the conterminous United States, *GCB Bioenergy*, 4, 509–520.

Millennium Ecosystem Assessment (2005) *Ecosystems and Human Well-being: Synthesis.* Island Press, Washington, DC.

Milner, S., Holland, R.A., Lovett, A. et al. (2015) Potential impacts on ecosystem services of land use transitions to second-generation bioenergy crops in GB, *GCB Bioenergy*, doi:10.1111/gcbb.12263.

Mishra, U., Torn, M. and Fingerman, K. (2013) *Miscanthus* biomass productivity within US croplands and its potential impact on soil organic carbon, *GCB Bioenergy*, 5, 391–399.

Monteith, J.L. (1977) Climate and the efficiency of crop production in Britain, *Philosophical Transactions of the Royal Society*, London, 281, 277–294.

Monti, A. (Ed.) (2013) *Switchgrass – A Valuable Biomass Crop for Energy.* Springer Verlag, London.

Mousdale, D.M. (2010) *Introduction to Biofuels.* CRC Press, Taylor and Francis Group, Boca Raton, FL.

Murphy, F., Devlin, G. and McDonell, K. (2013) *Miscanthus* production and processing in Ireland: An analysis of energy requirements and environmental impacts, *Renewable and Sustainable Energy Reviews*, 23, 412–420.

Nair, S.S., Kang, S., Zhang, X., Kang, S. et al. (2012) Bioenergy crop models: Descriptions, data requirements, and future challenges, *GCB Bioenergy*, 4, 620–633.

Naidu, S., Moose, S.P., Al-Shoaibi, A.K. et al. (2003) Cold tolerance of C_4 photosynthesis in *Miscanthus x giganteus*: Adaptation in amount and sequence of C_4 photosynthetic enzymes, *Plant Physiology*, 132, 1688–1697.

Naidu, S.L. and Long, S.P. (2004) Potential mechanism of low-temperature tolerance of C_4 photosynthesis in *Miscanthus x giganteus*: An in vivo analysis, *Planta*, 220, 145–155.

Nguyen, T.L.N. and Hermansen, J.E. (2015) Life cycle environmental performance of *Miscanthus* gasification versus other technologies for electricity production, *Sustainable Energy Technologies and Assessments*, 9, 81–94.

Nunn, C., St. John Hastings, A., Kalinina, O. et al. (2017) Environmental influences on the growing season duration and ripening of diverse *Miscanthus* germplasm grown in six countries, *Frontiers in Plant Science*, doi:10. 3389/fpls.2017.00907.

OECD (2008) *Economic Assessment of Biofuel Support Policies.* Organisation for Economic Development and Cooperation (OECD), Paris, France.

O'Flynn, M.G., Finnan, J.M., Curley, E.M., McDonnell, K.P. (2018) Effect of harvest time and soil moisture content on compaction, growth and harvest yield in a *Miscanthus* cropping system. *Agriculture*, doi:10.3390/agriculture8100148.

Panagopoulos, Y., Gassman, P.W., Kling, C.L. et al. (2017) Water quality assessment of large-scale bioenergy cropping scenarios for the Upper Mississippi and Ohio-Tennessee river basins, *Journal of the American Water Resources Association*, 53, 1355–1367.

Parrish, D.J. and Fike, J.H. (2005) The biology and agronomy of switchgrass for biofuels, *Critical Reviews in Plant Science*, 24, 423–445.

Paris Agreement (2015) Paris Agreement, FCCC/CP/2015/L.9/Rev.1.

Parikka, M. (2004) Global biomass fuel resources, *Biomass and Bioenergy*, 27, 613–620.

Pejchar, L. and Mooney, H.A. (2009) Invasive species, ecosystem services and human wellbeing, *Trends in Ecology and Evolution*, 24, 497–504.

Paulrud, S. and Laitila, T. (2010) Farmers' attitudes about growing energy crops: A choice experiment approach, *Biomass and Bioenergy*, 34, 1770–1779.

Peláez-Samaniego, M.R., Garcia-Perez, M., Cortez, L.B. et al. (2008) Improvements of Brazilian carbonization industry as part of the creation of a global biomass economy, *Renewable and Sustainable Energy Reviews*, 12, 1063–1086.

Perez-Harguindeguy, N., Diaz, S., Garnier, E. et al. (2013) New handbook for standardised measurement of plant functional traits worldwide, *Australian Journal of Botany*, http://dx.doi.org/10.1071/BT12225.

Perlack, R.D., Wright, L.L. and Turhollow, A.F. (2005) Biomass as a feedstock for a bioenergy and bioproducts industry: The technical feasibility of a billion-ton annual supply, *U.S. Department of Energy*, http://www.osti.gov/bridge.

Popp, J., Lakner, Z., Harangi-Rakos, M. et al. (2014) The effect of bioenergy expansion: Food, energy, and environment, *Renewable and Sustainable Energy Reviews*, 32, 559–578.

Power, A.G. (2010) Ecosystem services and agriculture: Tradeoffs and synergies, *Philosophical Transactions of the Royal Society*, B. 365, 2959–2971.

Prado, J.R., Segers, G., Voelker, T. et al. (2014) Biotech crop development: From idea to product, *Annual Review of Plant Biology*, 65, 769–790.

Price, L., Bullard, M., Lyons, H. et al. (2004) Identifying the yield potential of *Miscanthus x giganteus*: An assessment of the spatial and temporal variability of M. x giganteus biomass productivity across England and Wales, *Biomass and Bioenergy*, 26, 3–13.

Purdy, S.J., Maddison, A.L., Nunn, C.P. et al. (2017) Could *Miscanthus* replace maize as the preferred substrate for anaerobic digestion in the United Kingdom? Future breeding strategies, *GCB Bioenergy*, 9, 1122–1139.

Quinn, L.D., Allen, D.J. and Stewart, J.R. (2010) Invasiveness potential of *Miscanthus sinensis*: Implications for bioenergy production in the United States, *GCB Bioenergy*, 2, 310–320.

Rasse, D.P., Budai, A., O'Toole, A. et al. (2017) Persistence in soil of *Miscanthus* biochar in laboratory and field conditions, *Plos One*, doi:10.1371/pone.0184383.

Ragauskas, A.J., Williams, C.K., Davison, B.H. et al. (2006) The path forward for biofuels and biomaterials, *Science*, 311, 484–489.

Rauscher, B. and Lewandowski, I. (2006) *Miscanthus* horse bedding compares well to alternatives. In: S. Barth et al. (Eds.) *Perennial Biomass Crops for a Resource Constrained World*. Springer, Switzerland.

Richard, T.L. (2010) Challenges in scaling up biofuels infrastructure, *Science*, 329, 793–796.

Richter, G.M., Richie, A.B. and Dailey, A.G. (2008) Is UK biofuel supply for *Miscanthus* water limited? *Soil Use and Management*, 24, 235–245.

Richter, G.M., Agostini, F., Barker, A. et al. (2016) Assessing on-farm productivity of *Miscanthus* crops combining soil mapping, yield modelling and remote sensing, *Biomass and Bioenergy*, 85, 252–261.

Rist, L., Ser Huay Lee, J. and Koh, L.P. (2009) Biofuels: Social benefits, *Science*, 326, 1344.

Robertson, A.D., Davies, C.A., Smith, P. et al. (2015) Modelling the carbon cycle of *Miscanthus* plantations: Existing models and the potential for their improvement, *GCB Bioenergy*, 7, 405–421.

Robson, P., Mos, M., Clifton-Brown, J. et al. (2012) Phenotypic variation in senescence in *Miscanthus*: Towards optimising biomass quality and quantity, *Bioenergy Research*, 5, 95–105.

Robson, P., Jensen, E., Hawkins, S. et al. (2013) Accelerating the domestication of a bioenergy crop: identifying and modelling morphological targets for sustainable yield increases in *Miscanthus*, *Journal of Experimental Biology*, 64, 14, 4143–4155.

Roth, B., Jones, M.B., Burke, J. et al. (2013) The effects of land-use change from grassland to *Miscanthus x giganteus* on soil N_2O emissions, *Land*, 2, 437–451.

Roth, B., Finnan, J.M., Jones, M.B. et al. (2015) Are the benefits of yield responses to nitrogen fertilizer application in the bioenergy crop *Miscanthus x giganteus* offset by increased soil emissions of nitrous oxide? *GCB Bioenergy*, 7, 145–152.

Samson, R., Mani, S., Boddy, R., et al., (2005) The potential of C$_4$ perennial grasses for developing a global bioheat industry, *Critical Reviews in Plant Science*, 24, 461–495.

Scarlat, N. and Dallemand, J-F. (2011) Recent developments of biofuel/bioenergy sustainability certification: A global overview, *Energy Policy*, 39, 1630–1649.

Scarlat, N., Dallemand, J-F., Montforti-Ferrario, F. and Nita, V. (2015) The role of biomass and bioenergy in a future bioeconomy: Policies and facts, *Environmental Development*, 15, 3–34.

Scauflaire, J., Gourgue, M., Foucart, G. et al. (2013) *Fusarium miscanthi* and other *Fusarium* species as causal agents of *Miscanthus x giganteus* rhizome rot, *European Journal of Plant Pathology*, 137, 1–3.

Schnitzler, A. and Essl, F. (2015) From horticulture and biofuel to invasion: The spread of *Miscanthus* taxa in the USA and Europe, *European Weed Research Society*, 55, 221–225.

Schueler, V., Fuss, S., Stekel, J.C. et al. (2016) Productivity ranges of sustainable biomass potentials from non-agricultural land, *Environmental Research Letters*, 11, 074026, doi:10.1088/1748–9326/11/17/070426.

Schulze, E-D., Korner, C., Law, B.E. et al. (2012) Large-scale bioenergy from additional harvest of forest biomass is neither sustainable nor greenhouse gas neutral, *GCB Bioenergy*, 4, 611–616.

Searchinger, T., Heimlich, R., Houghton, R.A. et al. (2008) Use of U.S. croplands for biofuels increases greenhouse gases through emissions from land-use change, *Science*, 319, 1238–1240.

Searle, S. and Malins, C. (2014) Will energy crop yields meet expectations? *Biomass and Bioenergy*, 65, 1–12.

Searle, S. and Malins, C. (2015) A reassessment of global bioenergy potential in 2050, *GCB Bioenergy*, 7, 328–336.

Service, R.A. (2010) Is there a road ahead for cellulosic ethanol? *Science*, 329, 784–785.

Shortall, O.K. (2013) "Marginal land" for energy crops: Exploring definitions and embedded assumptions, *Energy Policy*, http://dx.doi.org/10.1016/j.enpol.2013.07.048.

Shrestha, P., Ibanez, A.B., Bauer, S. et al. (2015) Fungi isolated from *Miscanthus* and sugarcane biomass conversion, fungal enzymes, and hydrolysis of plant cell wall polymers, *Biotechnology for Biofuels*, doi:10.1186/s13068-o15-0221-3.

Sims, R.E.H., Hastings, A., Schlamadinger, B. et al. (2006) Energy crops: Current status and future prospects, *Global Change Biology*, 12, 2054–2076.

Slade, R., Bauen, A. and Gross, R. (2014) Global bioenergy resources, *Nature, Climate Change*, 4, 99–105.

Slusarkiewicz-Jarzina, A., Ponitka, A., Cerazy-Waliszewska, J. et al. (2017) Effective and simple in vitro regeneration system of *Miscanthus sinensis, M. giganteus* and *M. sacchariflorus,* for planting and biotechnology purposes, *Biomass and Bioenergy*, 107, 219–226.

Smeets, E.M.W., Lewandowski, I.M. and Faaij, A.P.C. (2009) The economical and environmental performance of *Miscanthus* and switchgrass production and supply chains in a European setting, *Renewable and Sustainable Energy Reviews*, 13, 1230–1245.

Smith, J. (2010) *Biofuels and the Globalisation of Risk: The Biggest Change in North-South Relationships since Colonialism?* Zed Books, London.

Smith, J.U., Gottschalk, P., Bellarby, J. et al. (2010) Estimating changes in Scottish soil carbon stocks using ECOSSE. I. Model description and uncertainties, *Climate Research*, 45, 122–179.

Smith, K. (Ed.) (2010) *Nitrous Oxide and Climate Change.* Earthscan, London.

Smith, L.J. and Torn, M.S. (2013) Ecological limits to terrestrial biological carbon dioxide removal, *Climatic Change*, 118, 89–103.

Smith, P. (2016) Soil carbon sequestration and biochar as negative emission technologies, *Global Change Biology*, 22, 1315–1324.

Smith, P., Bustamante, M., Ahammad, H. et al. (2014) Agriculture, forestry and other land uses (AFOLU). In: *Climate Change 2014: Mitigation of Climate Change. Contribution of Working Group III to the Fifth assessment Report of the IPCC*, pp. 870–886. Cambridge University Press, Cambridge and New York.

Smith, P., Davis, S.J., Creutzig, F. et al. (2016) Biophysical and economic limits to negative CO2 emissions, *Nature, Climate Change*, 6, 42–50.

Soman, C., Keymer, D.P. and Kent, A.D. (2018) Edaphic correlates of feedstock-associated diazotroph communities, *GCB Bioenergy*, doi:10.1111/gebb.12502.

Somerville, C., Youngs, H., Taylor, C. et al. (2010) Feedstocks for lignocellulosic biofuels, *Science*, 329, 790–792.

Staffas, L., Gustavsson, M. and McCormick, K. (2013) Strategies and policies for the bioeconomy and bio-based economy: An analysis of official national approaches, *Sustainability*, 5, 2751–2769.

Stampfl, P.F., Clifton-Brown, J.C. and Jones, M. (2007) Europe-wide GIS-based modelling system for quantifying the feedstock from *Miscanthus* and the potential contribution to renewable energy targets, *Global Change Biology*, 13, 2283–2295.

Styles, D. and Jones, M.B. (2007) Energy crops in Ireland: An assessment of their potential contribution to sustainable agriculture, electricity and heat production, *ERTDI Report Series* No. 70. Environmental Protection Agency, Ireland.

Styles, D. and Jones, M.B. (2008) *Miscanthus* and willow production – An effective land-use strategy for greenhouse gas emission avoidance in Ireland? *Energy Policy*, 36, 97–107.

Styles, D., Gibbons, J., Arwel, P. et al. (2015) Consequential life cycle assessment of biogas, biofuel and biomass energy options within an arable crop rotation, *GCB Bioenergy*, 7, 1305–1320.

Thomas, M.A., Ahiablame, L.M., Engel, B.A. and Chaubey, I. (2014) Modelling water quality impacts of growing corn, switchgrass, and *Miscanthus* on marginal soils, *Journal of Water Resources and Protection*, 6, 1352–1368.

Tilman, D., Socolow, R., Foley, A. et al. (2009) Beneficial biofuels – The food, energy, and environment trilema, *Science*, 325, 270–271.

Tonini, D., Hamelin, L., Wenzel, H. et al. (2012) Bioenergy production from perennial energy crops: A consequential LCA of 12 bioenergy scenarios including land-use changes, *Environmental Science & Technology*, 46, 13521–13530.

UNFCCC (2015) *United Nations Framework Convention on Climate Change*. Adoption of the Paris Agreement FCCC/CP/2015/10/Add.1.

USDA (2011) Planting and managing giant *Miscanthus* as a biomass crop, *Technical Note No.4* https://www.nrcs.usda.gov/Internet/FSE_DOCUMENTS/stelprdb1044768.pdf.

US DOE (2005) Biomass as feedstock for a bioenergy and bioproductivity industry: The technical feasibility of a billion-ton annual supply. http://www.eere.energy.gov/biomass/biomass_feedstocks.html.

US DOE (2006) Breaking the biological barriers to cellulosic ethanol: A joint research agenda. DOE/SC-0095, *US Department of Energy Office of Science and Office of Energy Efficiency and Renewable Energy* (www.doegenomestolife.org/biofuels/).

US DOE (2016) 2016 Billion-Ton Report: Advancing Domestic Resources for a Thriving Bioeconomy, *Volume 1: Economic Availability of Feedstocks*. M.H. Langholtz, B.J. Stokes and L.M. Eaton (Leads), ORNL/TM-2016/160. Oak Ridge National Laboratory, Oak Ridge, CA, TN. 448p, doi:10.2172/1271651. http://energy.gov/eere/bioenergy/2016-billion-ton-report.

Valentine, J. and Clifton-Brown, J. (2012) Food vs fuel: The use of land for lignocellulosic 'next generation' energy crops that minimize competition with primary food production, *GCB Bioenergy*, 4, 1–19, doi:10.1111/j.1757–1707.2011.01111.x.

Van der Weijde, T., Kiezel, A., Iqbal, Y. et al. (2016) Evaluation of *Miscanthus sinensis* biomass quality as feedstock for conversion into different bioenergy products, *GCB Bioenergy*, doi:10.1111/gebb.12355.

Van der Weijde, T., Huxley, L.M., Hawkins, S. et al. (2017) Impact of drought stress on growth and quality of *Miscanthus* for biofuel production, *GCB Bioenergy*, 9, 770–782.

Van Esbroeck, G.A., Hussey, M.A. and Sanderson, M.A. (2003) Variation between Alamo and Cave-in-Rock Switchgrass in response to photoperiod extension, *Crop Science*, 43, 639–643.

VanLoocke, A., Bernacchi, C.J. and Twine, T.E. (2010) The impact of *Miscanthus x giganteus* production on the Midwest US hydrological cycle, *GCB Bioenergy*, 2, 180–191.

Vaughan, N.E., Goough, C., Mander, S. et al. (2018) Evaluating the use of biomass energy with carbon capture and storage in low emission scenarios, *Environmental Research Letters*, 13, 044014.

Vermerris, W. (2008) *Genetic Improvement of Bioenergy Crops*. Springer, New York.

Villamil, M.B., Silvis, A.H. and Bollero, G.A. (2008) Potential *Miscanthus* adoption in Illinois: Information needs and preferred information channels, *Biomass and Bioenergy*, 32, 1338–1348.

Visser, P. and Pignatelli, V. (2001) Utilisation of *Miscanthus*. In: Jones, M.B. and Walsh, M. (Eds.) *Miscanthus for Energy and Fibre*. James & James, London.

Voigt, T.B. (2015) Are the environmental benefits of *Miscanthus x Giganteus* suggested by early studies of this crop supported by the broader and longer-term contemporary studies? *Global Change Biology: Bioenergy*, 7, 4, 567–569. https://doi.org/10.1111/gcbb.12150.

Wagner, M., Kiesel, A., Hastings, A., et al. (2017) Novel *Miscanthus* germplasm-based value chains: A life cycle assessment, *Frontiers of Plant Science*, doi:10.3389/fpls.2017.00980.

Wagner, M. and Lewandowski, I. (2017) Relevance of environmental impact categories for perennial biomass production, *GCB Bioenergy*, 9, 215–228.

Wang., Yamada and Kong, T. (2011) Establishment of an efficient in vitro culture and particle bombardment-mediated transformation systems in *Miscanthus sinensis Anderss.*, a potential bioenergy crop, *GCB Bioenergy*, 3, 322–332.

Wang, S., Wang, S., Lovett, A. et al. (2014) Significant contribution of energy crops to heat and electricity needs in Great Britain to 2050, *Bioenergy Research*, 7, 919–926.

WBGU (2008) Future Bioenergy and sustainable Land, online at www.wbgu.de.

Welfle, A., Gilbert, P. and Thornley, P. (2014) Securing a bioenergy future without imports, *Energy Policy*, 68, 1–14.

Whitaker, J., Field, J.L., Bernacchi, C.J. et al. (2018) Consensus, uncertainties and challenges for perennial bioenergy crops and land use, *GCB Bioenergy*, 10, 150–164.

Witzel, C-P. and Finger, R. (2016) Economic evaluation of *Miscanthus* production: A review, *Renewable and Sustainable Energy Reviews*, 53, 681–696.

Xue, S., Kalinina, O. and Lewandowski, I. (2015) Present and future options for *Miscanthus* propagation and establishment, *Renewable and Sustainable Reviews*, 49, 1233–1246.

Xue, S., Lewandowski, I. and Kalinina, O. (2017) *Miscanthus* establishment and management on permanent grassland in southwest Germany, *Industrial Crops and Products*, 108, 572–582.

Yan, J., Chen, W.L., Luo, F. et al. (2012) Variability and adaptability of *Miscanthus* species evaluated for energy crop domestication, *GCB Bioenergy*, 4, 49–60.

Yoder, J.R., Alexander, C., Ivanic, R. et al. (2015) Risks versus reward, a financial analysis of alternative contract specifications for *Miscanthus* lignocellulosic supply chain, *Bioenergy Research*, 8, 644–656.

Yost, M.A., Randall, B., Kitchen, N.R. et al. (2017) Yield potential and nitrogen requirement of *Miscanthus giganteus* on eroded soil, *Agronomy Journal*, 109, 684–695.

Yuan, J.S., Tiller, K.H., Al-Ahmed, H. et al., (2008) Plants to power: Bioenergy to fuel the future. *Trends in Plant Science*, 13, 421–429.

Yue, D., You, F. and Snyder, S.W. (2014) Biomass-to-bioenergy and biofuel supply chain optimisation: Overview, key issues and challenges, *Computer and Chemical Engineering*, 66, 36–56.

Zang, H., Blagodatskaya, E., Wen, Y. et al. (2018) Carbon sequestration and turnover in soil under the energy crop *Miscanthus*: Repeated ^{13}C natural abundance approach and literature synthesis, *GCB Bioenergy*, 10, 262–271.

Zhu, X.-G., Long, S.P. and Ort, D.R. (2008) What is the maximum efficiency with which photosynthesis can convert solar energy into biomass? *Current Opinion in Biotechnology*, 19, 153–159.

Zhuang, Q., Qin, Z. and Chen, M. (2013) Biofuel, land and water: Maize, switchgrass or Miscanthus? *Environmental Research Letters*, doi:10.1088/1748–9326/8/1/015020.

Zimmermann, J., Dauber, J. and Jones, M.B. (2012) Soil carbon sequestration during the establishment phase of *Miscanthus x giganteus*: A regional-scale study on commercial farms using ^{13}C natural abundance, *GCB Bioenergy*, 4, 453–461.

Zimmerman, J., Styles, D., Hastings, A. et al. (2014) Assessing the impact of within crop heterogeneity ('patchiness') in young *Miscanthus x giganteus* fields on economic feasibility and soil carbon sequestration, *GCB Bioenergy*, 6, 566–576, doi:10.1111/gebb.12084.

Zub, H.W. and Brancourt-Hulmel, M. (2010) Agronomic and physiological performances of different species of *Miscanthus*, a major energy crop: A review, *Agronomy for Sustainable Development*, 30, 201–214.

Zub, H.W., Arnoult, S. and Brancourt-Hulmel, M. (2011) Key traits for biomass production identified in different *Miscanthus* species at two harvest dates, *Biomass and Bioenergy*, 35, 637–651.

Index

Page numbers in *italics* and **bold** indicate Figures and Tables, respectively.